LEAD W
HARNESSIN
EXCEPTIC

THE POWER OF MENTOR

UNLEASHING POTENTIAL,
EMBRACING DIVERSITY AND
SHAPING EXTRAORDINARY
LEADERS

VOLUME - II

SREEKANTH GANESHI

Books By This Author

The Ultimate Leadership in You

The Power of Mentor – Volume I

"LEADERSHIP IS NOT ABOUT BEING THE BEST. LEADERSHIP IS ABOUT MAKING EVERYONE ELSE BETTER."

Copyright © Sreekanth Ganeshi 2023

All Rights Reserved.

ISBN: 9798859797042

This book has been published with all reasonable efforts taken to make the material error-free after the consent of the author. No part of this book shall be used or reproduced in any manner whatsoever without written permission from the author, except in the case of brief quotations embodied in critical articles and reviews.

The views expressed in this book are solely those of the author and should not be considered as expert instructions or commands. Each reader is responsible for their own actions and decisions.

The Author of this book is solely responsible and liable for its content including but not limited to the views, representations, descriptions, statements, information, opinions, and references ["Content"]. The Content of this book shall not constitute or be construed or deemed to reflect the opinion or expression of the Publisher or Editor. Neither the Publisher nor Editor endorse or approve the Content of this book or guarantee the reliability, accuracy or completeness of the content published herein and do not make any representations or warranties of any kind, express or implied, including but not limited to the implied warranties of merchantability, fitness for a particular purpose. The Publisher and Editor shall not be liable whatsoever for any errors, or omissions, whether such errors or omissions result from negligence, accident, or any other cause or claims for loss or damages of any kind, including without limitation, indirect or consequential loss or damage arising out of use, inability to use, or about the reliability, accuracy or sufficiency of the information contained in this book.

DEDICATION

This book is a heartfelt tribute to my beloved wife, Seema Ganeshi. Her unwavering support and boundless encouragement have been the driving force behind its creation. Throughout this journey, she has stood by my side, making countless sacrifices and unwavering compromises to ensure the success of this endeavour. Without her by my side, this book would not have come to fruition. Seema, your love, and dedication have been the ultimate inspiration, and I am forever grateful for your presence in my life.

Table of Contents

DEDICATION i

ACKNOWLEDGEMENTS v

PREFACE vii

INTRODUCTION 1

CHAPTER 7: OVERCOMING MENTORSHIP CHALLENGES: HOW TO NAVIGATE CONFLICTS AND OVERCOME OBSTACLES IN MENTORSHIP 5

CHAPTER 8: THE POWER OF REVERSE MENTORING: HOW LEARNING FROM A YOUNGER MENTOR CAN BENEFIT YOU 29

CHAPTER 9: BUILDING A STRONGER NETWORK: HOW MENTORSHIP CAN HELP YOU EXPAND YOUR PROFESSIONAL AND PERSONAL CONNECTIONS 56

CHAPTER 10: THE ROLE OF MENTORS IN LEADERSHIP DEVELOPMENT: HOW MENTORSHIP CAN HELP YOU BECOME A BETTER LEADER 76

CHAPTER 11: THE IMPACT OF TECHNOLOGY ON MENTORSHIP: HOW TO LEVERAGE DIGITAL TOOLS AND PLATFORMS FOR EFFECTIVE MENTORSHIP 99

CHAPTER 12: MENTORSHIP AND DIVERSITY: HOW TO FIND AND ENGAGE MENTORS FROM DIFFERENT BACKGROUNDS AND PERSPECTIVES 130

CHAPTER 13: MENTORSHIP BEYOND BOUNDARIES: HOW TO ESTABLISH AND MAINTAIN LONG-DISTANCE MENTOR-MENTEE RELATIONSHIPS 149

May I ask you for a small favor? 171

ACKNOWLEDGEMENTS

This book is an embodiment of the profound love and encouragement bestowed upon me by my cherished family and friends. First and foremost, I humbly express my gratitude to the divine power that guided me throughout this journey and led me to the place I stand today. Thank you, God, for granting me the strength and inspiration to pen these words.

To my exceptional **parents**, your unwavering support has been the cornerstone of my success. I am forever indebted to you for instilling in me the drive to achieve greatness. My heartfelt thanks to my younger sisters, **Jyothi and Aruna,** and my brother **Kiran Kumar,** whose sacrifices have paved the way for my triumphs. It is through the unwavering support of my family that I have realized my dreams.

A special note of appreciation goes to my soulmate, **Seema Ganeshi**. Your unyielding support in this literary endeavour has been nothing short of remarkable. Your belief in my aspirations has been the fuel that powered this journey.

I am also deeply grateful to my mentors, **Bijay Kumar Khandal and Geetika Khandal,** for their transformative guidance that reshaped my thinking. A heartfelt thank you to my The Power of Gratitude mentor, **Shobha Devraj,** for her wisdom and encouragement. Furthermore, I extend my sincere appreciation to **Som Bathla,** my book publishing mentor, whose expertise guided me through the process of writing and publishing. Also, my career banding mentor **Sakshi Chandraakar.**

Lastly, I extend my heartfelt thanks to you, dear reader. Your decision to pick up and peruse this book is a testament to your passion for personal growth and development. It is my earnest hope that these pages inspire and empower you on your journey of transformational leadership and mentorship.

Thank you from the depths of my heart for being a part of this profound adventure. Your support and engagement are the true sources of motivation that keep the essence of this book alive.

PREFACE

In the pages that follow, you will embark on an extraordinary odyssey through the realm of mentorship - a journey enriched by meticulous research, heartfelt testimonials, and real-world experiences. "The Power of Mentor" is not merely a compilation of theories; it is a profound exploration of the human spirit, illuminated by the transformative force of mentorship.

Drawing from a wealth of knowledge and expertise, this book brings to light the importance of mentors and their unparalleled ability to reshape destinies. Our quest for understanding has led us to dive deep into the realms of neuroscience and psychology, revealing the intricate workings of mentorship and its profound effects on the human brain and behaviour.

But beyond the scientific foundation, we have sought to capture the essence of mentorship through the poignant stories of those who have been touched by its magic. Real-world experiences and testimonials from individuals across diverse fields have breathed life into these pages, vividly illustrating how mentors have guided them to new heights, igniting the spark of inspiration that forever changed their lives.

The voices of these mentors and mentees resonate throughout the book, sharing the invaluable lessons and wisdom that have shaped their journeys. Through these authentic narratives, we bear witness to the extraordinary power of mentorship in fostering personal and professional growth, transforming mere aspirations into resounding achievements.

As you traverse the chapters of this book, you will discover practical insights into finding the right mentor, establishing meaningful connections, and cultivating trust and rapport. Through vivid case studies and inspiring examples, you will be inspired by the indelible impact mentors have had on celebrities, corporate leaders, educators, athletes, and those making a difference in their communities through non-profit initiatives.

We recognize that mentorship is a dynamic relationship, and the wisdom shared here extends not only to those seeking mentors but also to those eager to become mentors themselves. Our exploration of the psychology of mentorship lays bares the intricacies of this symbiotic connection, enriching both mentor and mentee, and shedding light on how to overcome challenges that may arise on this shared journey.

This book also delves into the vital role of technology in modern mentorship, bridging gaps and transcending boundaries to connect mentors and mentees from all corners of the world. It embraces the importance of diversity, understanding that unique perspectives fuel innovation and creating a tapestry of inspiration that knows no bounds.

"The Power of Mentor" is more than just words on paper; it is a testament to the profound impact mentorship can have on our lives and the lives of those we touch. It is a celebration of the indomitable human spirit and the potential we can unlock when guided by the wisdom and care of a mentor.

With each turn of the page, we invite you to embrace the transformative power of mentorship and to be inspired by the journeys of those who have walked this path before. May the wisdom shared within these pages ignite a passion for mentorship and may the stories of mentor-mentee relationships kindle a desire to create a better world, one connection at a time.

Welcome to "The Power of Mentor" - a journey that will inspire, uplift, and empower you to reach new heights and unleash the greatness within.

x

INTRODUCTION

Recap of The Power of Mentor Volume – I

Congratulations on completing the reading of Volume I and it's time to choose **"The Power of Mentor" Volume II!** This transformational book will equip you with invaluable insights and strategies to maximize the benefits of mentorship in your personal and professional life. Let's explore the potent chapters together:

Chapter 1: The Importance of Mentors: How They Can Transform Your Life

In this chapter, you'll understand why everyone needs a mentor. Discover the benefits of mentorship that will help you stay focused, motivated, and accountable for your goals. Explore the science of mentorship and how it positively impacts your brain, behavior, and overall well-being. Embrace the power of role models and learn why they have such a significant influence on our lives. Understand the pivotal role mentors play in your personal and professional growth, providing you with guidance, support, and opportunities for development. Lastly, identify the different types of mentors and figure out which one aligns best with your aspirations and goals.

Chapter 2: Finding the Right Mentor: How to Identify and Connect with the Best Mentors for You

Learn to define your goals and objectives, which will align your mentorship journey with your aspirations.

Build a strong network by actively engaging with potential mentors through various channels. Discover the qualities to look for in a mentor and how to choose the right one for your personal and professional growth. Take the initiative to approach your potential mentor and establish a strong foundation for your mentor-mentee relationship based on trust and rapport.

Chapter 3: Learning from Mentors: Strategies and Techniques for Maximizing the Benefits of Mentorship

Set clear expectations for your mentorship journey and communicate effectively with your mentor to achieve your goals. Hone your active listening skills to extract valuable insights and advice from your mentor's experiences. Learn how to take action on the knowledge gained from your mentor and apply it to enhance your personal and professional development. Be prepared to overcome challenges that might arise in your mentorship journey and learn valuable techniques to navigate obstacles effectively.

Chapter 4: Mentors in the Real World: Case Studies and Examples of Successful Mentor-Mentee Relationships

Explore real-world examples of successful mentor-mentee relationships in different domains. Learn from celebrity mentors, corporate leaders, educators, coaches, and non-profit mentors making a positive impact in their communities. These case studies will inspire and motivate you to see the profound impact of mentorship on personal and professional growth.

Chapter 5: Becoming a Mentor: How to Pay It Forward and Help Others Achieve Their Goals

Discover the benefits of being a mentor and how it can be a fulfilling journey of personal growth. Understand the qualities of a good mentor and what it takes to be an effective mentor for others. Learn how to identify and connect with potential mentees and provide the necessary support and feedback to guide them towards success. Finally, learn the art of gracefully concluding a mentorship relationship once the mentee's goals are achieved.

Chapter 6: The Psychology of Mentorship: Understanding the Dynamics of Mentor-Mentee Relationships

Delve into the psychology of mentorship and understand the psychological benefits it brings to both mentees and mentors. Gain insights into the mentor-mentee power dynamic and learn how to navigate relationship patterns effectively. Appreciate the role of trust in mentorship and identify common barriers that can hinder the effectiveness of mentorship relationships.

"The Power of Mentor" Volume I is a comprehensive guide that empowers you to make the most out of your mentorship journey. Embrace the knowledge, strategies, and case studies to become a transformational leader and positively impact your life and those around you. Get ready to embark on a journey of growth, success, and empowerment through the power of mentorship!

CHAPTER 7: OVERCOMING MENTORSHIP CHALLENGES: HOW TO NAVIGATE CONFLICTS AND OVERCOME OBSTACLES IN MENTORSHIP

"Through open communication and a commitment to growth, mentorship challenges can be transformed into valuable opportunities for learning and development."
– John C Maxwell

Mentorship is an essential component of personal and professional development. A mentor provides guidance, support, and encouragement to a mentee, helping them to achieve their goals and reach their potential. However, mentorship can sometimes face challenges that can make it difficult for both the mentor and the mentee. In this response, we will explore some common challenges that can arise in mentorship and how to navigate them.

1. **Misaligned Expectations:** Misaligned expectations can create conflict between the mentor and the mentee. The mentor may expect the mentee to follow a specific plan, while the mentee may have different ideas about what they want to achieve. To overcome

this challenge, it is essential to establish clear expectations and goals from the outset of the mentorship relationship. The mentor and mentee should discuss and agree on what they hope to achieve from the mentorship and what steps they will take to reach these goals.

2. **Lack of Communication:** Effective communication is crucial in any mentorship relationship. When communication breaks down, misunderstandings and conflicts can arise. To overcome this challenge, the mentor and mentee should establish a regular communication schedule and discuss any concerns openly and honestly. It is also essential to actively listen to each other and be willing to provide constructive feedback.

3. **Personality Conflicts:** Personality conflicts can arise when the mentor and mentee have different working tyles or personalities. This can make it difficult to work together effectively. To overcome this challenge, it is important to acknowledge and accept each other's differences. The mentor and mentee should focus on their shared goals and work together to find a way to accommodate each other's working styles.

4. **Busy Schedules:** Busy schedules can make it challenging for the mentor and mentee to find time to work together. To overcome this challenge, the mentor and mentee should establish a regular meeting schedule that works for both parties. They should also

prioritize their mentorship relationship and make a commitment to dedicate time to work together.

5. **Lack of Progress:** A lack of progress can be frustrating for both the mentor and mentee. To overcome this challenge, it is important to assess the reasons why progress is not being made. The mentor and mentee should work together to identify any obstacles and develop a plan to overcome them. They should also celebrate small successes along the way to maintain motivation and momentum.

Common Conflicts in Mentorship Relationships and How to Address Them

In mentorship relationships, conflicts can arise despite efforts to align expectations, maintain communication, manage personalities, ensure accountability, navigate busy schedules, and drive progress. As a leadership mentor, it's crucial to address these conflicts head-on. This chapter provides detailed solutions for common conflicts that may occur in mentorship relationships, focusing on aspects beyond the points mentioned earlier.

1. **Differing Learning Styles:**

- Identify learning preferences: Recognize that mentees have unique learning styles. Some may prefer visual aids, while others thrive through hands-on experiences or verbal explanations. Encourage mentees to reflect on their learning preferences and share them with you.

- Adapt mentoring strategies: Tailor your mentoring approach to accommodate the mentee's learning style. Provide resources, activities, or examples that align with their preferred learning methods. Flexibility in your approach will enhance the mentee's engagement and understanding.

- Foster continuous feedback: Regularly check in with the mentee to assess how well the mentoring strategies are working for them Encourage them to share their experiences and provide suggestions on how you can better support their learning style.

2. **Power Imbalance:**

- Encourage open dialogue: Acknowledge the power dynamics that exist in a mentorship relationship. Create a safe space for the mentee to express concerns or voice disagreements. Promote an environment where both parties feel comfortable discussing their perspectives openly.

- Facilitate equal participation: Actively seek input from the mentee during discussions and decision-making process. Ensure that their thoughts, ideas and opinions are valued and given due consideration. Empower the mentee to contribute as an equal partner in the mentorship.

- Share experiences of vulnerability: Share your own experiences of vulnerability and challenges to demonstrate that you are approachable and have faced similar situations. This helps to humanize the mentorship dynamic and fosters a more balance relationship.

3. **Lack of Networking Opportunities:**

- Expand the mentee's network: Introduce the mentee to relevant contact from your professional network. Facilitate networking opportunities through industry events, conferences, or online communities. Encourage the mentee to actively engage with these connections to broaden their professional circle.

- Share resources and connections: Offer to share articles, books or online platforms that can enhance the mentee's networking skills and knowledge. Provide guidance on how to initiate and maintain professional relationships.

- Role-play networking scenarios: Conduct role-playing exercises to help the mentee practice networking skills. Offer constructive feedback on their approach, communication style and ability to establish meaningful connections.

As a leadership mentor addressing conflicts in mentorship relationships goes beyond aligning expectations, communication, personalities, accountability, schedules and progress. By tailoring your mentoring strategies to different learning styles, fostering open dialogue to balance power dynamics, and providing networking opportunities, you can resolve conflicts and create a more enriching mentorship experience. Effective conflict resolution leads to stronger mentor-mentee relationships and greater personal and professional growth.

Strategies For Dealing with Difficult Mentors or Mentees

- **Knowing when to listen and provide guidance:** To handle difficult mentor or mentee dynamics, it's crucial to recognize the appropriate moments to actively listen and offer guidance. Understanding the balance between the two enables effective communication and support. For example, if your mentee is facing a personal challenge, such as a conflict with a colleague, it's important to lend a listening ear and provide empathy. On the other hand, when your mentee seeks advice on career development, you can offer guidance based on your expertise and experiences.

- **Set a positive example that others can follow:** As a leader, it's essential to embody the qualities you expect from your mentor or mentee. By demonstrating integrity, resilience, and professionalism, you inspire others to follow suit, fostering a productive mentorship environment. For instance, If you emphasize the importance of integrity and transparency, ensure that your own actions align with these values. By consistently displaying positive behaviors, you inspire others to emulate them, fostering a culture of excellence.

- **Find common interests and values with your mentee:** Building a strong connection with your mentee requires identifying shared interests and values. Establishing common ground creates a foundation for understanding, trust, and collaboration, enhancing the effectiveness of mentorship. For example, if both you and your mentee are passionate about community service, you can collaborate on a volunteer project together. Finding common ground creates a sense of camaraderie and helps forge a deeper understanding and trust between mentor and mentee.

- **Recognize and utilize your own strengths:** By acknowledging your own unique strengths, you can leverage them to support and guide your mentor or mentee effectively. Utilizing your expertise and experiences allows you to provide valuable insights and contribute to their growth. For instance, if you excel in strategic thinking, you can provide valuable insights and help your mentee develop their strategic planning abilities. By utilizing your strengths, you enhance the mentorship experience and contribute to their growth.

- **Give clear and straightforward advice:** When offering advice and feedback, clarity is paramount. Articulate your suggestions in a straightforward manner, ensuring that your mentor or mentee comprehends the guidance provided. Avoid ambiguity to facilitate their development. For example, if your mentee seeks guidance on improving their presentation skills, you can provide specific suggestions on areas they can work on, such as using visual aids effectively or practicing confident body language. Clear and straightforward advice enables your mentee to understand and implement your recommendations more effectively.

- **Help expand your mentee's network and open doors when appropriate:** As a powerful mentor, you can help broaden your mentee's professional network by introducing them to relevant contacts and resources. By opening doors and creating opportunities, you empower their growth and increase their chances of success. For instance, if you know someone in your industry who can provide valuable insights or potential collaborations for your mentee, you can facilitate introductions. By expanding their network, you empower their growth and increase their chances of success.

The Importance of Clear Communication in Mentorship Relationships

Mentoring proves invaluable for mentors and mentees

alike, representing a bond founded on trust, respect, and communication. Effective communication plays a vital role in mentoring relationships by establishing explicit expectations, fostering rapport, and encouraging personal and professional development.

- **Establish clear expectations:** It is vital to set clear goals and objectives for the mentoring relationship. As a mentor, clearly define the frequency and mode of communication, the scope of the relationship, and the desired outcomes. By establishing clear expectations, both parties can stay focused and aligned on their goals.

- **Practice active listening:** Active listening is a fundamental skill in mentoring relationships. As a mentor, provide undivided attention to your mentee, avoiding distractions. Listen attentively, seeking to understand their perspective. Ask open-ended questions to clarify their thoughts and feelings.

- **Offer constructive feedback:** Feedback is essential for growth. As a mentor, provide constructive feedback that helps your mentee improve their skills and achieve their goals, specific, timely, and focus on behaviors or performance, rather than personal characteristics.

- **Utilize effective communication tools:** Take advantage of modern communication tools to enhance your mentoring relationship. Email, instant messaging, and video conferencing are examples of tools that can facilitate communication. Select the appropriate mode for each situation. For instance, use email for detailed information and video conferencing for complex discussions.

- **Respect boundaries:** Respect your mentee's time and availability. Communicate withing agreed-upon times and channels. Be mindful of their personal and professional boundaries, avoiding topics that may be uncomfortable or inappropriate.

Addressing Imbalances in Mentorship Relationships

Addressing imbalances in mentoring relationships is crucial for fostering a fair and effective learning experience. This point presents powerful and simple strategies that empower mentors and mentees to create a balanced and mutually beneficial mentorship dynamic.

1. **Foster Transparent Communication:** Establish an environment of open and honest communication where mentors and mentees can discuss imbalances openly. Encourage both parties to express their concerns, expectations, and aspirations. Transparent conversations enable the identification of areas requiring adjustment and lay the foundation for balanced mentorship relationships.

2. **Emphasize mutual growth:** Shift the focus from a one-sided mentor-to-mentee dynamic to partnership centered on mutual growth. Encourage mentors to recognize the valuable insights and perspectives mentees bring to the table. Likewise, empower mentees to share their unique experiences and ideas. This fosters an environment of equality and promotes a more balanced mentorship experience.

3. **Set clear goals and boundaries:** Establish clear goals and expectations for the mentorship relationship. Ensure both mentor and mentee have a shared understanding of their roles, responsibilities, and boundaries. By defining these parameters upfront, imbalances can be identified early on and addressed proactively.

4. **Encourage Feedback and Evaluation:** Regularly seek feedback from mentors and mentees to assess the effectiveness of the mentorship relationship. Create a culture that values constructive feedback. Actively encourage mentees to provide feedback to mentors, and vice versa. This feedback loop allows for ongoing evaluation and adjustment, helping to address imbalances as they arise.

5. **Provide Mentor Training and Resources:** Equip mentors with the necessary skills and resources to navigate imbalances in mentorship relationships. Offer training programs or resources that address bias, cultural differences, and power dynamics. By enhancing mentors' awareness and understanding,

they can better support mentees and contribute to a more balanced and inclusive mentorship experience.

6. **Foster Diversity and Inclusion:** Promote diversity and inclusion in mentorship programs by actively seeking mentors and mentees from diverse backgrounds. Encourage cross-cultural exchanges and expose mentees to different perspectives. This creates an inclusive and equitable mentorship environment, minimizing imbalances based on factors such as gender, race, or ethnicity.

7. **Regularly Assess and Adjust:** Continuously assess the mentorship relationship and be open to making necessary adjustments. If imbalances persist or new ones emerge, take proactive steps to address them. Regular check-ins and evaluations provide opportunities for course correction, ensuring the mentorship relationship remains balanced and beneficial for all parties involved.

How to Manage Mentorship Relationships When Goals are Priorities Shift

Living in a constantly connected world presents challenges in maintaining focus and effectiveness as a leader. The multitude of distractions can divert our attention and hinder our ability to lead with power and confidence. This point explores the importance of prioritizing and staying focused amidst distractions, with a particular emphasis on mentoring relationships.

- **Many distractions lead to unfocused efforts**: In today's always-on society, distractions abound, making it difficult to concentrate on essential tasks. The temptation to engage with devices and multitask can hinder our ability to be fully present and attentive. A personal story highlights the struggle of balancing attention between family and external distractions, shedding light on the importance of focused engagement.

- **Priorities in mentoring:** Setting priorities within mentoring relationships is crucial for effective growth and development. Without clear priorities, both mentors and mentees may become overwhelmed and struggle to achieve their goals. This section emphasizes the need to establish goals and prioritize them within the mentoring journey.

- **Identify your goals:** Mentees and mentors should identify and articulate their respective goals for the mentoring relationship. By recording these aspirations and leveraging mentoring software, individuals can track their ideas effectively, setting a strong foundation for their collaboration.

- **Weight your goals:** To avoid overwhelming themselves and their mentoring partners, mentees need to assess and prioritize their goals realistically.

Assigning numeric or qualitative values to each goal helps determine their relative importance and feasibility. Mentors play a vital role in offering guidance and feedback to help mentees make informed decisions about prioritization.

- **Refine your list:** After weighing their goals, mentees should revisit their list and make necessary adjustments based on feedback and personal evaluation. A revised and ordered list ensures that the most important goals receive primary attention. It is essential for mentors to support their mentees throughout this process and provide honest input when certain goals may require specialized expertise.

- **Take action:** With a refined list of prioritized goals, both mentors and mentees are ready to take action. This section emphasizes that mentoring is an ongoing process that requires continuous learning and adaptability. By focusing on current goals and regularly reassessing their progress, individuals can embark on a fulfilling and lifelong journey of development.

The Role of Boundaries in Mentorship Relationships

Establishing clear boundaries is essential for creating a healthy and effective mentoring relationship. Boundaries provide a framework for mutual respect, trust, and professional growth. This moment explores the role of boundaries in mentorship relationships and highlights their significance in fostering a positive mentoring experience.

- **Defining Boundaries in Mentorship**: Boundaries can be defined as guidelines and limits that determine the scope and nature of the mentorship relationship. These boundaries serve to create a safe and respectful environment for both the mentor and the mentee. Clear communication and understanding of boundaries from the outset lay the foundation for a successful mentorship journey.

Boundaries for Mentors

- **Professional Boundaries:** Mentors must establish professional boundaries to maintain a productive and ethical mentorship relationship. This involves maintaining confidentiality, respecting personal boundaries and avoiding conflicts of interest. By upholding these boundaries, mentors create a sense of trust and professionalism with their mentees.

- **Time and availability boundaries:** Mentors need to set realistic expectations regarding their time and availability. Clearly communicating their preferred methods of communication, response times, and availability for meetings helps manage mentees' expectations and prevents undue stress or disappointment. Establishing boundaries around their time ensures mentors can provide quality guidance and support.

- **Expertise and scope boundaries:** Mentors should define the scope of their expertise and be transparent about their limitations. This helps mentees understand the areas where the mentor can provide valuable guidance and identify potential gaps that may require additional resources or support. By acknowledging their boundaries, mentors foster trust and avoid providing inaccurate or misleading advice.

Boundaries For Mentees
- **Respect for Mentor's Time and Expertise:** Mentees should respect their mentor's time commitments and adhere to agreed-upon schedules and deadlines. Being mindful of the mentor's expertise and not overstepping boundaries by seeking assistance beyond the mentor's area of specialization demonstrates respect for the mentor's limitations.

- **Active Engagement and Accountability:** Mentees play an active role in the mentorship relationship and should take responsibility for their own growth. This includes being prepared for meetings, following through on agreed-upon actions, and respecting the mentor's guidance and feedback. By demonstrating accountability, mentees show their commitment and value the mentor's time and expertise.

Nurturing Boundaries in Mentorship
- **Open communication and feedback:** Maintaining healthy boundaries requires open and honest communication between mentors and mentees.

Regular check-ins and feedback sessions provide an opportunity to discuss any concerns, clarify expectations, and address potential boundary-related issues. This fosters mutual understanding and strengthens the mentorship relationship.

- **Flexibility and adaptability:** Boundaries may need to be adjusted as the mentorship relationship evolves. Both mentors and mentees should be open to reevaluating and modifying boundaries when necessary. This flexibility ensures that the mentorship remains effective and aligned with the evolving needs and goals of both parties.

KEY TAKEAWAYS

CHAPTER 7: OVERCOMING MENTORSHIP CHALLENGES – HOW TO NAVIGATE CONFLICTS AND OVERCOME OBSTACLES IN MENTORSHIP

Common Conflicts in Mentorship Relationships and How to Address Them

Mentorship is vital for personal and professional development, but it can face obstacles. Let's explore common challenges and their solutions.

- **Misaligned Expectations:** Establish clear goals and expectations from the start. Discuss and agree on what both mentor and mentee hope to achieve and the steps to reach those goals.

- **Lack of Communication:** Maintain regular communication and openly discuss concerns. Actively listen and provide constructive feedback to foster effective communication.

- **Personality Conflicts:** Accept and acknowledge differences in working styles. Focus on shared goals and find ways to accommodate each other's preferences.

- **Busy Schedules:** Establish a regular meeting schedule that works for both mentor and mentee.

Prioritize the mentorship and dedicate time to work together.

- **Lack of Progress:** Assess reasons for a lack of progress. Identify obstacles and develop a plan to overcome them. Celebrate small successes to maintain motivation.

Additional Conflicts and Solutions:
1. **Differing Learning Styles:** Identify mentees' learning preferences and adapt mentoring strategies accordingly. Foster continuous feedback to support their learning style.

2. **Power Imbalance:** Encourage open dialogue and equal participation. Share experiences of vulnerability to create a balanced relationship.

3. **Lack of Networking Opportunities:** Expand the mentee's network by introducing relevant contacts. Share resources and connections to enhance networking skills. Conduct role-playing exercises to practice networking scenarios.

Strategies for Dealing with Difficult Mentors or Mentees

1. Know when to listen and guide: Recognize when to listen attentively and when to offer guidance, fostering effective communication and support.

2. Set a positive example: Demonstrate integrity, resilience, and professionalism to inspire others, creating a productive mentorship environment.

3. Find common interests and values: Identify shared interests and values to build a strong connection, enhancing understanding and collaboration.

4. Utilize your own strengths: Leverage your expertise and experiences to support and guide your mentor or mentee effectively contributing to their growth.

5. Give clear and straightforward advice: Provide specific guidance in a straightforward manner to facilitate comprehension and development.

6. Expand their network and create opportunities: Introduce your mentee to relevant contacts and resources, opening doors for their professional growth and success.

The Importance of Clear Communication in Mentorship Relationships

1. Establish clear expectations: Define goals, frequency of communication, and desired outcomes to ensure both mentor and mentee are aligned and focused.

2. Practice active listening: Give undivided attention, listen attentively, and ask open-ended questions to understand your mentee's perspective.

3. Offer constructive feedback: Provide specific and timely feedback focused on behaviors or performance to help your mentee grow and achieve their goals.

4. Utilize effective communication tools: Take advantage of modern tools like email, instant messaging, and video conferencing to enhance communication based on the situation.

5. Respect boundaries: Be mindful of your mentee's time and availability, communicate within agreed-upon times and channels and avoid discussing uncomfortable or inappropriate topics.

Addressing Imbalances in Mentorship Relationships

1. Foster transparent communication: Encourage open and honest conversations to discuss imbalances and adjust accordingly.

2. Emphasize mutual growth: Shift to a partnership mindset that values the insights and experiences of both mentors and mentees.

3. Set Clear Goals and Boundaries: Establish shared expectations, roles, responsibilities, and boundaries from the beginning.

4. Encourage feedback and evaluation: Regularly seek feedback to assess and address imbalances as they arise.

5. Provide mentor training and resources: Equip mentors with skills to navigate bias, cultural differences, and power dynamics.

6. Foster diversity and inclusion: Promote diversity in mentorship programs, encouraging cross-cultural exchanges and exposure to different perspectives.

7. Regularly assess and adjust: Continuously evaluate the mentorship relationship, making necessary adjustments to maintain balance.

How to Manage Mentorship Relationships When Goals are Priorities Shift

1. Acknowledge distractions: In a world full of distractions, recognize the importance of staying focused and present in mentoring relationships.

2. Set clear goals: Define specific goals for the mentoring relationship and articulate them together. This provides a roadmap for prioritization.

3. Identify and weigh goals: Mentees should assess and assign values to their goals, considering their importance and feasibility. Mentors can offer guidance in this process.

4. Refine and adjust: Continuously evaluate and refine the list of goals based on feedback and personal evaluation. Prioritize the most important ones.

5. Take Action: With a prioritized list, both mentors and mentees can take action. Emphasize that mentoring is an ongoing process requiring adaptability and continuous learning.

The Role of Boundaries in Mentorship Relationships

1. Defining Boundaries: Boundaries are guidelines that determine the scope and nature of mentorship. They create a safe and respectful environment for mentors and mentees.

Boundaries for Mentors:
- Professional Boundaries: Maintain confidentiality, respect personal boundaries, and avoid conflicts of interest.

- Time and availability boundaries: Communicate preferred methods of communication response times, and availability to manage expectations.

   Expertise and scope boundaries: Define areas of expertise and be transparent about limitations.

Boundaries for Mentees:
- Respect for Mentor's time and expertise: Adhere to agreed-upon schedules and deadlines and seek assistance within the mentor's area of specialization.

- Active engagement and accountability: Be prepared for meetings, follow through on actions and respect the mentor's guidance and feedback.

Nurturing Boundaries in Mentorship:
- Open communication and feedback: Regularly discuss concerns, clarify expectations, and address boundary-related issues.

- Flexibility and adaptability: Adjust boundaries as the mentorship relationship evolves to meet changing needs and goals.

CHAPTER 8: THE POWER OF REVERSE MENTORING: HOW LEARNING FROM A YOUNGER MENTOR CAN BENEFIT YOU

> *Before you are a leader, success is all about growing yourself. When you become a leader, success is all about growing others* – Jack Welch

It is important to recognize the power of reverse mentoring and how learning from a younger mentor can benefit you. Reverse mentoring refers to the practice of pairing older, more experienced professionals with younger, less experienced individuals to foster knowledge exchange and learning. Here are some clear details on how reverse mentoring can be beneficial.

1. **Fresh Perspective:** Younger mentors bring a fresh and innovative outlook on industry, technology, and trends. They have grown up in the digital era and possess a deep understanding of emerging technologies and social media platforms. Engaging with a younger mentor can help you gain new perspectives, challenge your assumptions, and think outside the box.

2. **Technical Literacy:** Today's younger generation has grown up with technology as an integral part of their lives. They are often well-versed in the latest tools, software, and digital platforms. By engaging with a younger mentor, you can enhance your technological literacy, learn new skills, and stay updated on the latest digital advancements. This knowledge can significantly benefit you in the workplace.

3. **Reverse Mentoring as a Learning Opportunity:** Reverse mentoring provides a unique learning opportunity where you can expand your knowledge in areas that you may not be familiar with. For example, if you are a seasoned leader in finance, your younger mentor could teach you about social media marketing or data analytics. Embracing this opportunity can help you become a more well-rounded leader.

4. **Bridging the Generation Gap:** Reverse mentoring promotes better understanding and collaboration between different generations in the workplace. It creates an inclusive environment where diverse perspectives are valued. By actively participating in reverse mentoring, you can bridge the generation gap, foster a sense of mutual respect, and create a more harmonious and productive work culture.

5. **Leadership Development:** Engaging with a younger mentor can help you refine your leadership skills. Younger professionals often possess a strong drive, ambition, and enthusiasm. Observing and learning from their leadership style can inspire you to adapt

your own approach, become more agile, and embrace change. Additionally, mentoring relationships, regardless of age, provide opportunities for personal growth, self-reflection, and the development of coaching and mentorship skills.

6. **Employee Engagement and Retention:** Implementing reverse mentoring programs can boost employee engagement and retention. It shows that you value the opinions and contributions of all employees, regardless of their age or experience level. This inclusive approach can create a positive work environment, increase job satisfaction, and foster loyalty among your team members.

7. **Embrace a Growth Mindset:** Adopting a growth mindset is the key to harnessing the power of reverse mentoring. Understand that learning can happen at any age or experience level. Embrace the idea that your younger mentor can offer valuable insights and knowledge that you can benefit from, regardless of your position or seniority.

8. **Active Listening and Curiosity:** To make reverse mentoring effective, practice active listening and curiosity. Create an environment where your younger mentor feels comfortable sharing their ideas, experiences and perspectives. Actively listen to their insights, ask open-ended questions, and encourage them to elaborate on their ideas. This will foster a deeper level of understanding and facilitate meaningful knowledge exchange.

9. **Set Clear Goals:** Establish clear goals and expectations for the reverse mentoring relationship. Outline specific areas where you would like to gain knowledge or skills from your younger mentor. By setting goals, both you and your mentor can align your efforts and work towards achieving tangible outcomes.

10. **Create a Safe and Inclusive Environment:** Ensure that the reverse mentoring is built on trust, respect, and inclusivity. Foster an environment where your younger mentor feels empowered to share their thoughts without fear of judgment. Emphasize that their contributions are valued and that their perspectives can have a significant impact on the organization.

11. **Be Open to Change and Adaptability:** Reverse mentoring can challenge existing beliefs, practices, and ways of thinking. As a powerful leader, be open to change and adaptable in your approach. Embrace the opportunity to learn new technologies, strategies, and perspectives that your younger mentor brings to the table. This adaptability will not only benefit you but also inspire your team to embrace change and innovation.

12. **Two-way Learning:** Reverse mentoring should be a mutually beneficial experience. While you can learn from your younger mentors, they can also benefit from your experience and wisdom. Encourage a two-way learning dynamic where both parties share their

knowledge and insights. This will create a synergistic relationship where everyone involved grows and develops.

13. **Apply Learned Knowledge:** Reverse mentoring is only effective if you apply the knowledge and insights gained from the relationship. Actively seek opportunities to implement the ideas and strategies discussed with your younger mentor. This will not only enhance your leadership skills but also demonstrate the value and impact of reverse mentoring within your organization.

To fully benefit from reverse mentoring, it is crucial to approach the relationship with an open mind, humility, and a genuine desire to learn. Actively listen to your younger mentor, ask thought-provoking questions, and be willing to step out of your comfort zone. By embracing the power of reverse mentoring, you can enhance your leadership skills, adapt to the changing business landscape, and drive innovation within your organization.

The Benefits of Reverse Mentorship for Older Professionals

1. **Stay relevant and up to date:** In a rapidly changing business landscape, staying relevant is crucial. Reverse mentoring allows older professionals to bridge the generational gap and stay up to date with

the latest trends, technologies, and industry insights. Younger mentors can share their knowledge of emerging tools, social media platforms, and digital strategies, helping older professionals adapt and thrive in a fast-paced environment.

2. **Gain fresh perspective:** Engaging with a younger mentor provides an opportunity for older professionals to gain fresh perspectives and challenge their own assumptions. Younger professionals often bring new ideas, innovative thinking, and a different worldview. By actively listening to their insights and experiences, older professionals can broaden their mindset, think creatively, and approach problem-solving from different angles.

3. **Enhance technological literacy:** Technology plays a significant role in today's workplace. Younger mentors, who have grown up in a digital era, can help now with software, navigating social media platforms, or leveraging digital tools for greater efficiency. This knowledge is valuable for older professionals looking to leverage technology in their roles and improve their overall productivity.

4. **Promote personal growth and development:** Reverse mentorship offers a unique opportunity for older professionals to continue their personal growth and development. Mentoring relationships fosters a sense of curiosity, self-reflection, and continuous learning. Engaging with a younger mentor can expose older professionals to new ideas, methodologies, and

career paths, helping them expand their skill set and pursue new opportunities.

5. **Develop cross-generational collaboration skills:** In today's diverse workforce, cross-generational collaboration is essential for success. By participating in reverse mentorship, older professionals can develop their collaboration skills and build effective relationships with younger colleagues. This inclusive approach encourages knowledge sharing, mutual respect and teamwork across different generations, fostering a positive work environment and driving overall team performance.

6. **Share wisdom and experience:** While older professionals can benefit from the insights of their younger mentors, they also have valuable wisdom and experience to share. Reverse mentorship creates a platform for older professionals to pass on their knowledge, expertise, and lessons learned throughout their careers. This exchange of wisdom can be highly valuable for the development of younger professionals and contribute to a stronger, more cohesive workforce.

7. **Boost engagement and retention:** Implementing reverse mentorship programs can boost employee engagement and retention, particularly among older professionals. It demonstrates that the organization values their contributions and seeks to invest in their growth and development. This sense of support and

engagement can increase job satisfaction, loyalty and overall retention rates within the company.

To fully reap the benefits of reverse mentorship, it's important for older professionals to approach the relationship with an open mind, humility, and a genuine willingness to learn from their younger mentors. Actively engage in conversations, ask for feedback and be receptive to new ideas. By embracing the power of reverse mentorship, older professionals can continue to grow, adapt, and contribute as effective and powerful leaders in their respective fields.

How to Find and Connect with Younger Mentors

1. **Leverage professional networks:** Reach out to your professional networks, both online and offline, to identify potential younger mentors. Attend industry conferences, seminars, and networking events where you are likely to meet younger professionals who can offer mentorship. Use platforms like LinkedIn to connect with individuals who align with your interests and expertise.

2. **Seek mentorship programs:** Many organizations and educational institutions offer mentorship programs that pair experienced professionals with younger individuals. Look for such programs in your industry or community and apply to become a mentor or mentee. These programs often provide a structured framework for establishing mentorship relationships.

3. **Engage with younger employees within your organization:** If you work in an organization that has a diverse workforce, take the opportunity to engage with younger employees. Attend team-building activities, social events, or cross-departmental projects where you can interact with individuals from different age groups. Building relationships within your own organization can lead to fruitful mentorship opportunities.

4. **Volunteer or participate in community activities:** Get involved in community activities or volunteer organizations that attract younger individuals. This can include professional associations, industry-specific groups, or nonprofit organizations. By actively participating in such activities, you can connect with younger professionals who share your interest and passions.

5. **Utilize online platforms:** Online platforms provide a wealth of opportunities to connect with younger mentors. Join relevant forums, social media groups, or online communities where discussions on industry-related topics take place. Actively contribute to these platforms and engage in conversations to build relationships with younger professionals.

6. **Attend seminars or workshops led by younger experts:** Keep an eye out for seminars workshops, or speaking events where younger experts are sharing their knowledge and insights. Attend these events where younger experts are sharing their knowledge

and insights. Attend these events and try to connect with the speakers afterwards. Express your interest in their expertise and inquire about the possibility of establishing a mentorship relationship.

7. **Be clear about your objectives:** Before approaching a potential younger mentor, clarify your objectives and what you hope to gain from the mentorship. Are you seeking guidance in a specific area of expertise? Do you want to learn about new technologies or industry trends? Clearly articulating your goals will help you find a suitable mentor who can meet your expectations.

8. **Approach with genuine interest and respect:** When reaching out to younger professionals, approach them with genuine interest and respect. Be clear about why you are seeking their mentorship and express your admiration for their expertise and achievements. Show that you value their time and insights and be open to reciprocating by offering your own knowledge and experience when relevant.

9. **Customize the mentorship relationship:** Once you have connected with a younger mentor customize the relationship to meet both of your needs. Discuss and establish boundaries, expectations, and communication channels. Some mentors may prefer in-person meetings, while others may be more comfortable with virtual communication. Tailor the mentorship experience to fit the preferences of both parties.

10. **Provide opportunities for mutual learning:** Remember that mentorship is a two-way street. While you can benefit from your younger mentor's insights, experiences, and expertise, offer opportunities for them to learn from you as well. Share your wisdom, experiences, and lessons learned throughout your career. Creating a mutually beneficial mentorship dynamic will enhance the overall relationship and foster a deeper level of learning.

11. **Maintain regular communication:** Consistent communication is key to building a strong mentorship relationship. Schedule regular meetings or check-ins to discuss progress, challenges, and goals. Stay engaged and provide updates on your own growth and development. Demonstrating your commitment and dedication to the mentorship will strengthen the bond with your younger mentor.

12. **Show gratitude and appreciation:** Throughout the mentorship journey, express gratitude and appreciation for the time, guidance, and support provided by your younger mentor. Acknowledge their contributions and the impact they have had on your personal and professional growth. A simple thank-you note or a small token of appreciation can go a long way in strengthening the mentorship relationship and showing your genuine gratitude.

13. **Continue learning from multiple sources:** While having a younger mentor is valuable, it's important to continue seeking knowledge from various sources.

Supplement your mentorship relationship by reading books, attending seminars, taking online courses, or engaging with other experienced professionals in your field. Embrace a growth mindset and continue to expand your knowledge and skills beyond mentorship.

14. **Evaluate and reassess the mentorship:** Regularly evaluate the progress and effectiveness of the mentorship relationship. Reflect on the goals you set at the beginning and assess whether they are being met. If adjustments or modifications are needed, communicate openly with your mentor, and discuss how the mentorship can be optimized to better serve your needs.

15. **Be open to new perspectives and feedback:** One of the key benefits of reverse mentoring is gaining new perspectives and receiving feedback. Embrace the opportunity to hear different viewpoints, challenge your own assumptions, and be open to constructive criticism. Use the insights and feedback provided by your younger mentor as opportunities for growth and self-improvement.

16. **Extent mentorship beyond professional topics:** While the focus of the mentorship may primarily be on professional growth, don't be afraid to explore personal topics as well. Building a deeper connection and understanding with your younger mentor can contribute to a more meaningful and holistic mentorship experience. Discuss personal

development, work-life balance, and other topics that can enhance your overall well-being.

17. **Share successes and celebrate milestones:** When you achieve milestones or experience successes as a result of your mentorship, share them with your younger mentor. Celebrate together and acknowledge the positive impact of mentorship on your journey. This not only strengthens the mentorship relationship but also reinforces the value of reverse mentoring as a powerful tool for growth and development.

Finally, remember that finding and connecting with younger mentors is a continuous process. As you progress in your career and interests evolve, new mentorship opportunities may arise. Stay proactive, be open to new connections, and continue seeking guidance and learning from younger mentors. By embracing the power of reverse mentoring, you can continuously expand your knowledge, challenge your perspectives, and become an even more effective and powerful leader.

Strategies for Leveraging the Skills and Knowledge of Younger Mentors

1. **Identifying their areas of expertise:** Take the time to understand the specific skills, knowledge, and experiences that your younger mentor possesses. Identify the areas in which they excel and align those with your own goals or areas where you seek

improvement. This will help you leverage their expertise effectively.

2. **Setting clear objectives:** Clearly define the objectives and desired outcomes you hope to achieve through the mentorship. Communicate these objectives to your younger mentor so that they have a clear understanding of how they can best support you. This will ensure that both parties are on the same page and working towards shared goals.

3. **Actively seeking their input:** Actively seek input and ideas from your younger mentor on relevant topics and challenges you are facing. Create a safe and open space for them to share their thoughts and perspectives. Encourage them to provide their insights and actively listen to their suggestions. Their fresh perspective and innovative thinking can provide valuable solutions and approaches that you may not have considered.

4. **Learning from their experiences:** Younger mentors often have experiences and perspectives that are different from your own. Take the opportunity to learn from their unique journeys and the challenges they have faced. By understanding their experiences, you can gain valuable insights and potentially avoid similar pitfalls or mistakes in your own endeavors.

5. **Encouraging collaboration and knowledge sharing:** Foster a collaborative environment where knowledge sharing is encouraged between you and

your younger mentor. Create opportunities for joint projects, brainstorming sessions, or discussions where both parties can contribute their expertise. By leveraging each other's strengths, you can achieve more impactful results and foster a sense of mutual growth.

6. **Embracing technology and digital skills:** Younger mentors often have a strong command of technology and digital tools. Embrace their knowledge and expertise in these areas. Allow them to introduce you to new tools. Software, or platforms that can enhance your productivity and effectiveness. By leveraging their digital skills, you can stay current and leverage technology to your advantage.

7. **Providing opportunities for leadership and growth:** Younger mentors may have aspirations for leadership and growth themselves. Provide them with opportunities to take on leadership roles or challenging projects where they can apply their skills and talents. By empowering them to grow and develop, you can further strengthen the mentorship relationship and create a symbiotic learning environment.

8. **Seeking feedback and continuous improvement:** Regularly seek feedback from your younger mentor on your performance, ideas, and strategies. Be open to constructive criticism and use it as an opportunity for growth and improvement. Their feedback can

provide valuable insights and help you refine your skills and approaches.

9. **Recognizing and celebrating their contributions:** Acknowledge and celebrate the contributions of your younger mentor. Give credit where it is due and publicly recognize their efforts and achievements. This not only validates their expertise but also fosters a positive mentorship dynamic where both parties feel values and appreciated.

10. **Paying it forward:** As you benefit from the skills and knowledge of your younger mentor consider paying it forward by mentoring others. Share your own experiences, insights, and expertise with those who may benefit from your guidance. By becoming a mentor yourself, you can contribute to the development of future leaders and create a culture of continuous learning within your organization.

Effective leveraging of younger mentors requires a collaborative and mutually beneficial approach. Be open, receptive, and appreciative of their contributions. By leveraging their skills and knowledge, you can enhance your own capabilities and achieve greater success as a leader.

Addressing Potential Power Imbalances in Reverse Mentorship Relationships

Reverse mentorship relationships, where a younger individual mentors an older professional, can sometimes come with potential power imbalances due to differences in age, experiences, and hierarchical positions. Addressing these imbalances is crucial to ensure a healthy and productive mentorship dynamic. Here are some strategies:

1. **Establishing a safe and inclusive environment:** Create a safe space where both mentor and mentee feel comfortable expressing their thoughts and ideas. Foster an environment that encourages open dialogue, active listening, and respect for diverse perspectives. This helps to minimize power differentials and promotes equal participation.

2. **Recognizing the value of generational perspectives:** Reverse mentorship is built on the premise that younger individuals bring unique insights and perspectives. Acknowledge the value of their experiences and knowledge, regardless of their hierarchical position. Emphasize the importance of mutual learning and growth, valuing the contributions of both parties.

3. **Embracing a learner mindset:** As the mentee in a reverse mentorship, adopt a learner mindset and be open to new ideas and approaches. Recognize that age or seniority does not equate to having all the answers. By embracing a learner mindset, you create a level playing field that allows for the exchange of knowledge and experiences.

4. **Encouraging open communication:** Encourage open and honest communication between the mentor

and mentee. Foster an environment where both parties feel comfortable sharing their thoughts, concerns, and aspirations. This helps to address power imbalances and ensures that the relationship is based on mutual trust and respect.

5. **Balancing power dynamics:** Actively work to balance power dynamics within the mentorship relationship. Encourage the mentee to take on leadership roles and provide opportunities for them to showcase their skills and expertise. This helps to create a more balanced exchange of knowledge and fosters a sense of empowerment for the younger mentor.

6. **Establishing clear boundaries and expectations:** Set clear boundaries and expectations for the mentorship relationship from the beginning. Discuss topics such as confidentiality, goals, and the frequency of meetings. This clarity ensures that both parties understand their roles and responsibilities, minimizing any potential misuse of power.

7. **Seeking feedback and reflection:** Regularly seek feedback from both mentor and mentee to evaluate the effectiveness of the relationship. Encourage open discussions about power dynamics and any concerns that arise. Engage in reflective practices to identify areas for improvement and ensure that the mentorship remains equitable and beneficial for both individuals.

In July-2022, in New Your City, USA a seasoned executive, Sarah, who had been working in the financial industry for over 30 years, recognized the need to bridge the generational gap and better understand emerging digital trends. She sought a reverse mentorship relationship to learn from a younger professional's perspective. Through her professional network, Sarah connected with Alex, a talented and tech-savvy analyst in her organization. Sarah and Alex started their mentorship journey by openly discussing their goals, expectations, and potential power imbalances. They acknowledge that Sarah's seniority could create an unintentional power dynamic. To address this, they established a safe and inclusive environment where both could freely share their expertise and insights. During their mentorship, Sarah actively embraced a learner mindset, recognizing the value of Alex's expertise in technology and digital trends. She respected Alex's opinions and encouraged open communication, ensuring that their discussions were a two-way exchange of knowledge. To balance power dynamics, Sarah empowered Alex to take the lead in certain areas, such as organizing workshops on emerging technologies for the team. By giving Alex opportunities to showcase his skills and expertise, Sarah ensured that the mentorship relationship was equitable and mutually beneficial.

Throughout their mentorship journey, Sarah regularly sought feedback from Alex to ensure that the power dynamics were balanced and that their relationship remained productive. They engaged in open discussions about any concerns or challenges they faced, addressing them promptly to maintain a healthy dynamic. Sarah and Alex also reflected on their mentorship experience together. They discussed the power dynamics they observed and identified strategies to further mitigate them. By openly acknowledging and addressing the potential imbalances, they created a strong foundation of trust and respect. As their mentorship progressed, Sarah realized that she could offer valuable insights and guidance to Alex as well. She shared her experiences navigating the financial industry and provided strategic advice on career growth. This reciprocal exchange of knowledge further strengthened their relationship and enhanced their overall learning experience.

By the end of their mentorship, Sarah and Alex had developed a strong bond and had both grown personally and professionally. Sarah gained valuable insights into digital trends and emerging technologies, while Alex benefited from Sarah's wisdom and industry expertise. Their successful mentorship journey showcased the importance of addressing potential power imbalances in reverse mentorship relationships. Through open communication, mutual respect, and a learner mindset, Sarah and Alex created an equitable and fruitful dynamic that transcended their age and experiences differences.

Common Challenges in Reverse Mentorship Relationships and How to Overcome Them

While reverse mentorship can be a rewarding experience, it's important to be aware of and address common challenges that may arise. Here are some powerful details on how to overcome these challenges:

1. **Resistance to Vulnerability:** Reverse mentorship required both parties to be open, vulnerable, and willing to challenge their own perspectives. However, older professionals may find it challenging to let go of their expertise and admit that they have things to learn from their younger mentors. This resistance to vulnerability can hinder the growth and effectiveness of the relationship.

   To overcome this challenge, it's important to create a safe and non-judgmental environment where both mentors and mentees feel comfortable sharing their thoughts, ideas, and vulnerabilities. Encourage open and honest communication and foster an atmosphere of trust and mutual respect. Remind yourself that being vulnerable is a strength and catalyst for personal and professional growth.

2. **Communication Styles:** Different generations often have distinct communication styles. Younger mentors may prefer digital communication channels, while older professionals may be more accustomed to in-person meetings. It's essential to find a balance that works for both parties. Agree on preferred modes of communication, such as face-to-face meetings, video calls, or instant messaging platforms, to ensure effective and comfortable communication.

3. **Knowledge and skill gaps:** Reverse mentorship is intended to bridge knowledge and skill gaps, but it's possible that the younger mentor may lack expertise in certain areas. To overcome this challenge, be clear about your expectations and goals for the mentorship. Identify specific areas where you want to learn from your younger mentor and collaborate on finding resources or additional mentors to supplement their knowledge.

4. **Stereotypes and Preconceptions:** Stereotypes and preconceptions based on age can sometimes affect reverse mentorship relationships. Older professionals may assume that younger mentors lack experience or wisdom, while younger mentors may believe that older professionals are resistant to change. To overcome these biases, approach the relationship with an open mind and challenge your own preconceptions. Focus on individual capabilities and expertise rather than making assumptions based on age.

5. **Time constraints:** Both mentors and mentees often have busy schedules and competing priorities: This can make it challenging to find dedicated time for mentorship activities. To address this, establish a regular meeting schedule that works for both parties. Prioritize the mentorship relationship and set aside dedicated time for discussions, knowledge sharing, and goal setting. Be flexible and willing to accommodate each other's commitments.

6. **Resistant to Change:** Reverse mentorship often involves embracing new technologies, methodologies, or perspectives. Resistance to change can hinder the effectiveness of the relationship. To overcome this challenge, adopt a growth mindset and embrace the opportunity to learn from your younger mentor. Be open to trying new approaches, technologies, or strategies, even if they challenge your comfort zone. Recognize that growth requires embracing change.

7. **Lack of Alignment:** Sometimes, the mentor and mentee may not have aligned goals or interests. This can lead to a lack of engagement and productivity. To address this, have open and honest conversations about your expectations, interests, and goals from the outset. Identify areas of common ground and explore how mentorship can benefit both parties. If the misalignment persists, be open to reevaluating the mentorship relationship and seeking alternative mentoring opportunities.

By proactively addressing these challenges, you can create a more effective and productive reverse mentorship relationship. Remember to maintain open communication, embrace diversity of thought, and approach the relationship with a willingness to learn and grow together.

KEY TAKEAWAYS

CHAPTER 8: THE POWER OF REVERSE MENTORING: HOW LEARNING FROM A YOUNGER MENTOR CAN BENEFIT YOU

The Benefits of Reverse Mentorship for Older Professionals

1. Reverse mentoring provides fresh perspectives, technical literacy, and learning opportunities from younger mentors.
2. It bridges the generation gap, fosters inclusivity, and promotes collaboration in the workplace.
3. Engaging with younger mentors refines leadership skills, boosts employee engagement, and enhances retention.
4. Adopt a growth mindset, practice active listening, set clear goals, and create a safe and inclusive environment for effective reverse mentoring.
5. Reverse mentorship helps older professionals stay relevant, gain fresh perspectives, enhance technological literacy, promote personal growth, and develop cross-generational collaboration skills.
6. Sharing wisdom and experience in reverse mentorship can benefit younger professionals and boost engagement and retention.
7. Approach reverse mentorship with an open mind, humility, and a genuine willingness to learn from younger mentors.

By embracing the power of reverse mentorship, leaders can enhance their skills, adapt to change, and foster a culture of innovation within their organization. Older professionals can stay relevant, gain fresh perspectives, and develop collaborative skills through reverse mentorship.

How to Find and Connect with Younger Mentors

1. Leverage professional networks and online platforms.
2. Seek mentorship programs and engage with younger employees in you organization.
3. Attending seminars or workshops led by younger experts.
4. Be clear about your objectives and approach with genuine interest and respect.
5. Customize the mentorship relationship and provide opportunities for mutual learning.
6. Maintain regular communication and show gratitude and appreciation.
7. Continue learning from multiple sources and be open to new perspectives and feedback.
8. Extend mentorship beyond professional topics and share successes and celebrate milestones.

To stay proactive, and open-minded, and continuously seek guidance and learning from younger mentors to enhance your own growth and development.

Strategies for Leveraging the Skills and Knowledge of Younger Mentors

1. Identify their areas of expertise and align them with your goals.
2. Set clear objectives and communicate them to your mentor.
3. Actively seek their input and encourage them to share their perspectives.
4. Learn from their unique experiences and challenges.
5. Foster collaboration and create opportunities for knowledge sharing.
6. Embrace their digital skills and leverage technology.
7. Provide opportunities for their leadership and growth.
8. Seek feedback and use it for continuous improvement.
9. Recognize and celebrate their contributions.
10. Pay it forward by mentoring others.

By implementing these strategies, you can maximize the benefits of the mentorship relationship and create a mutually beneficial learning environment.

Addressing Potential Power Imbalances in Reverse Mentorship Relationships

1. Create a safe and inclusive environment for open dialogue and respect.
2. Value the unique perspectives of younger mentors.
3. Embrace a learner mindset, regardless of age or seniority.

4. Foster open communication based on trust and mutual respect.
5. Balance power dynamics by empowering the mentee.
6. Set clear boundaries and expectations from the beginning.
7. Seek feedback and engage in reflection for continuous improvement.

Common Challenges in Reverse Mentorship Relationships and How to Overcome Them

1. Create a Safe and non-judgmental environment for vulnerability.
2. Find a balance in communication styles that works for both parties.
3. Identify and address knowledge and skill gaps through clear expectations and supplemental resources.
4. Challenge stereotypes and preconceptions based on age.
5. Prioritize dedicated time for mentorship activities despite busy schedules.
6. Embrace change and a growth mindset to overcome resistance.
7. Ensure alignment of goals and interests for a productive relationship.

By addressing these challenges head-on, you can foster a successful and fulfilling reverse mentorship experience.

CHAPTER 9: BUILDING A STRONGER NETWORK: HOW MENTORSHIP CAN HELP YOU EXPAND YOUR PROFESSIONAL AND PERSONAL CONNECTIONS

> *"Leaders should influence others in such a way that it builds people up, encourages and edifies them so they can duplicate the attitude in others"* – Bob Goshen

Mentorship is indeed a valuable tool for expanding both professional and personal networks. Building a strong network is essential for personal growth, career advancement, and accessing opportunities. Mentors provide guidance, support, and a wealth of knowledge and experience, helping mentees develop their skills and expand their connections. Here's how mentorship can assist you in expanding your professional and personal networks.

- **Access to diverse network:** Mentors often have extensive professional networks built over the years. By engaging with a mentor, you gain access to their contacts and connections, expanding your own network. They can introduce you to professionals in your field, industry leaders, and other influential individuals. These connections can open doors to new opportunities, collaborations, and partnerships.

- **Guidance in relationship-building:** Mentors can offer valuable insights and guidance on how to effectively network and build meaningful relationships. They can teach you networking strategies, provide introductions, and help you navigate professional events. Their experience and advice can accelerate your networking efforts and make them more impactful.

- **Recommendations and endorsements:** As a mentor get to know your skills, strengths, and aspirations, they can become advocates for your professional growth. They may recommend you for job opportunities, introduce you to potential employers or clients, or endorse your skills to others. These recommendations carry weight and can significantly enhance your chances of expanding your network and securing valuable connections.

- **Personalized introductions and referrals:** Mentors can make personalized introductions to individuals who align with your goals or interests. They can facilitate networking meetings, arrange informational interviews, or refer you to specific contacts. These targeted connections are often more effective than general networking efforts, as they are tailored to your needs and can lead to deeper, more meaningful connections.

- **Building confidence and credibility:** Working closely with a mentor can boost your confidence and credibility. As you develop new skills and gain valuable insights, you become more self-assured in networking situations. A confident and knowledgeable mentee is more likely to make a positive impression and establish stronger connections, which can further expand their network.

- **Long-term support and sponsorship:** A mentor-mentee relationship often extends beyond immediate networking opportunities. Mentors can provide ongoing support and guidance throughout your career journey, acting as sponsors who actively advocate for your professional development. They may include you in their own networks., provide ongoing advice and help you navigate challenges. This sustained support contributes to the expansion of your network over time.

Mentorship is a two-way street. While mentors can provide significant support, it's essential to actively engage and contribute to the relationship. Show appreciation for your mentor's guidance, be proactive in seeking their advice and continuously work on strengthening your skills and knowledge. By leveraging mentorship effectively, you can tap into a wealth of resources, expand your network, and accelerate your personal and professional growth.

How Mentorship Can Help You Build a More Diverse Network

Mentorship plays a crucial role in helping individuals build a more diverse network. By actively engaging in a mentor-mentee relationship, you can benefit from the following aspects that contribute to fostering diversity within your network.

- **Different perspectives and experiences:** Mentors often come from diverse backgrounds, cultures, and experiences. By connecting with mentors who have different worldviews or have faced unique challenges, you gain exposure to perspectives that you may not have encountered otherwise. This exposure broadens your understanding and helps you build empathy, adaptability, and inclusivity in your network.

- **Introductions to diverse professionals:** Mentors can introduce you to professionals from various backgrounds, ethnicities, genders, and industries. They can help you form connections with individuals who bring different skills, expertise, and perspectives to the table. This diversity within your network fosters creativity, innovation, and the ability to navigate diverse work environments.

- **Opportunities for cross-cultural learning:** A mentor-mentee relationship provides an opportunity for cross-cultural learning. Your mentor can share insights about their cultural norms, values, and customs, helping you develop cultural intelligence. This knowledge enables you to navigate diverse settings with sensitivity, connect with individuals from different backgrounds and build more inclusive relationships.

- **Access to diverse communities and resources:** Mentors can connect you with diverse communities, professional organizations, and resources that cater to underrepresented groups. These communities often provide networking events, conferences, and platforms where you can engage with a diverse range of professionals. Actively participating in these spaces can expose you to new perspectives, amplify your voice, and provide opportunities for collaboration.

- **Challenging biases and fostering inclusion:** A mentor can play a crucial role in challenging your biases and expanding your understanding of diversity and inclusion. They can help you recognize unconscious biases and support your efforts to address them. By actively engaging with your mentor's guidance, you can develop a more inclusive mindset and create a network that embraces diversity in all its forms.

- **Expanding opportunities and perspectives:** A diverse network brings forth a range of opportunities and perspectives. By connecting with professionals from different backgrounds, industries and experiences, you open doors to new opportunities for collaboration, innovation, and growth. Diverse perspectives can also enhance problem-solving and decision-making processes, leading to more robust outcomes in your personal and professional endeavors.

It's essential to seek mentorship opportunities with individuals who value diversity and inclusion. Look for mentors who actively engage in activities that promote diversity belong to diverse networks themselves or have demonstrated a commitment to fostering inclusivity. By actively participating in a mentor-mentee relationship with a focus on diversity, you can expand your network, increase your cultural competence, and contribute to creating a more inclusive professional and personal sphere.

Strategies For Finding Mentors Outside of Your Immediate Network

We explore the profound impact of mentorship in expanding our professional and personal connections. Finding mentors outside of our immediate network is a powerful way to broaden our perspectives, gain invaluable insights, and forge meaningful relationships. Let me share with you the powerful story of Sarah, whose journey encompasses all the strategies for finding mentors outside of your immediate network.

Sarah was a young marketing professional with big dreams of making a significant impact in her industry. Despite her passion and dedication, she realized that her immediate network didn't provide the mentorship she desired to accelerate her growth. Undeterred, she embarked on a journey to find mentors outside of her immediate circle. Embracing the power of curiosity, Sarah attended industry conferences, seminars, and networking events. She approached each interaction with genuine interest, eager to learn from experienced professionals. During one such conference, she struck up a conversation with a keynote speaker, expressing her admiration for their work. Little did she know that this encounter would mark the beginning of a transformative mentorship journey. Recognizing the abundance of knowledge beyond her immediate network, Sarah cultivated a mindset of limitless possibilities. She understood that mentors were waiting to be discovered in unexpected places. Through the power of technology, Sarah expanded her reach. She joined online communities and engaged with professionals who shared her passion for marketing. It was through one of these platforms that she connected with a seasoned marketing executive who had achieved remarkable success in the industry.

Sarah sought out specialized communities and organizations aligned with her interests. She became an active participant in marketing associations, attending workshops and seminars. In one of these events, she had the opportunity to listen to a panel of marketing experts. Inspired by their insights, she approached one of the panellists after the session and shared her aspirations. The panellist was impressed by Sarah's enthusiasm and introduced her to a network of accomplished marketers who eventually became her mentors. Harnessing the power of introductions, Sarah reached out to her existing connections. She shared her aspirations for finding a mentor and asked if anyone knew of professionals who could guide her. A close friend introduced Sarah to a successful marketing consultant, emphasizing her dedication and thirst for knowledge. The introduction led to a powerful mentorship relationship that propelled Sarah's career to new heights. Throughout her journey, Sarah encountered setbacks and faced moments of doubt. However, her unwavering motivation and persistence fueled her search for mentors outside of her immediate network. She realized that building a stronger network through mentorship required active engagement, continuous learning, and a willingness to step outside her comfort zone.

Sarah's story exemplifies the transformative power of finding mentors outside of your immediate network. By embracing curiosity, cultivating an abundance mindset, leveraging technology, seeking specialized communities, and harnessing the power of introductions, Sarah expanded her professional and personal connections exponentially. The mentors she found along the way provided invaluable guidance, share their wisdom, and opened doors to new opportunities. As you embark on your own journey to build a stronger network, let Sarah's story serve as a powerful reminder that mentors are waiting to be discovered in unexpected places. Stay curious, believe in the abundance of possibilities, leverage technology to expand your reach, immerse yourself in specialized communities and embrace the power of introductions. With unwavering motivation and persistence, you too can find mentors who will propel your growth, broaden your horizons, and transform your professional and personal life.

Leveraging Mentorship to Build Relationships with Other Professionals in Your Field

Sarah's inspiring journey, she discovered that mentorship not only expanded her network but also served as a powerful tool to build relationships with other professionals in her field. The guidance and support she received from her mentors opened doors to a broader network of like-minded individuals who shared her aspirations and passion for marketing. As Sarah's mentorship relationships grew stronger, she began to realize the immense value of connecting with her mentors' networks. They introduced her to industry leaders, influential decision-makers, and successful entrepreneurs, providing her with opportunities to forge meaningful relationships. Sarah understood that building relationships with other professionals in her field was a vital aspect of her growth and success. Leveraging the mentorship, she received. Sarah actively engaged with her mentors' connections. She attended networking events, business conferences, and industry gatherings where she had the chance to meet these professionals face-to-face. By leveraging her mentors' credibility and support, she was able to establish genuine connections with other influential individuals. During one networking event, Sarah approached a prominent marketing executive who was introduced to her by one of her mentors. She expressed her admiration for the executive's accomplishments and shared her journey as a mentee. Impressed by Sarah's dedication and the recommendation from her mentor, the executive took a keen interest in Sarah's development. This encounter marked the beginning of a valuable professional relationship that would later lead to collaboration on a groundbreaking marketing campaign.

Sarah also understood the power of reciprocity in building relationships. She actively sought opportunities to contribute to the success of her mentors and their connections. She shared her knowledge, expertise and resources whenever possible, always striving to add value to the lives and careers of those she met through mentorship. This commitment to giving back further solidified her reputation and fostered deeper connections with professionals in her field. As Sarah continued to leverage mentorship to build relationships, she realized the importance of nurturing these connections over time. She maintained regular communications with her mentor and other professionals, offering updates on her progress, seeking advice, and expressing gratitude for their guidance. These ongoing interactions allowed Sarah to deepen her relationships, gain further insights, and create a network of trusted colleagues and friends. Through the combined power of mentorship and relationship building, Sarah's network expanded exponentially. She not only gained knowledge and support from her mentors but also forged alliances with other professionals who shared her passion and vision. Collaborative opportunities emerged joint ventures were formed, and her influence within the marketing community grew stronger.

Sarah's story is a testament to the transformative impact of leveraging mentorship to build relationships with other professionals in your field. By actively engaging with mentors' networks. Attending networking events, contributing to the success of others and nurturing these connections over time, Sarah created a network that was not only vast but also filled with individuals who believed in her and were invested in her success.

As you embark on you own journey of mentorship remember that the relationships you cultivate along the way are as important as the mentorship itself. Be proactive in leveraging the guidance and support of your mentors to connect with other professionals in your field. Attend industry events, engage in meaningful conversations, and give back whenever possible. By leveraging mentorship to build relationships, you will expand your network, open doors to new opportunities, and create a community of like-minded individuals who will support and inspire you on your path to success.

How Mentorship Can Help You Build Meaningful Personal Connections

On Sarah's remarkable journey, she discovered that mentorship not only had a profound impact on her professional growth but also played a pivotal role in building meaningful personal connections. Through the guidance and support of her mentors, Sarah experienced the transformative power of mentorship in shaping not only her career but also her personal life. As Sarah delved deeper into her mentorship relationships, she realized that her mentors genuinely cared about her holistic development. They recognized the importance of not only nurturing her professional skills but also fostering her personal growth and well-being. They shared their own life experiences, offered advice on work-life balance, and provided insights on personal development. Inspired by her mentors, Sarah began to prioritize self-reflection and personal growth. She sought their guidance on developing a positive mindset, managing stress, and finding fulfilment in both her personal and professional endeavors. Through their mentorship, Sarah learned valuable life lessons that extended far beyond her career aspirations.

Sarah also discovered that mentorship opened doors to meaningful personal connections with like-minded individuals who shared her values and aspirations. As she engaged with her mentors and their networks, she connected with individuals who not only provided professional guidance but also became lifelong friends and confidants. During one of her mentorship sessions. Sarah shared her aspirations for creating a positive impact in her community. To her surprise, her mentor introduced her to a group of social entrepreneurs who were passionate about making a difference. Through their shared commitment to creating social change, Sarah build deep personal connections with individuals who shared her vision and became her partners in various community initiatives. Furthermore, Sarah realized that the support and encouragement she received from her mentors had a ripple effect on her personal relationships. The personal growth she experienced through mentorship transformed her interactions with family, friends and loved ones. She became more compassionate, empathetic and attentive to the needs of others. Her newfound confidence and self-assuredness radiated in all areas of her life, strengthening her personal connections, and fostering deeper relationships.

Sarah understood that mentorship not only provided guidance in her professional journey but also enriched her personal life in immeasurable ways. The bond she shared with her mentors extended beyond business matters, as they became trusted advisors and friends. Their mentorship helped her navigate challenges, make important life decisions, and stay true to her values. As you on board on your own mentorship journey, remember that mentorship has the power to create profound personal connections. Embrace the guidance and support of your mentors not only in your professional endeavors but also in your personal growth. Seek their wisdom on life's challenges, foster connections with like-minded individuals, and let the mentorship journey share your personal relationships positively. Sarah's story exemplifies the transformative potential of mentorship to build meaningful personal connections. By embracing the holistic guidance of her mentors, she not only achieved professional success but also cultivated a network of individuals who shared her values, passion, and dreams. Let her journey inspire you to seek mentorship as a means to not only advance your career but also nurture personal growth and foster deep and meaningful connections that will enrich every aspect of your life.

The Importance of Networking Etiquette in Mentorship Relationships

During Sarah's incredible journey, she discovered the vital role that networking etiquette played in her mentorship relationships. As she deepened her connections and fostered meaningful relationships with her mentors and other professionals, Sarah understood that maintaining proper networking etiquette was crucial for nurturing and sustaining these valuable connections. Sarah recognized that networking etiquette went beyond mere formalities; it was a reflection of her professionalism, respect, and gratitude towards her mentors and colleagues. She understood that practising good networking etiquette not only enhanced her reputation but also strengthened the bond with her mentors, making them more willing to invest their time and wisdom in her growth.

First and foremost, Sarah prioritized active and attentive listening during her interactions. She recognized that mentors were sharing their valuable insights and experiences, and she made it a point to fully engage in the conversation. Sarah asked thoughtful questions, sought clarification when needed and genuinely absorbed the wisdom being shared. This demonstrated her respect for her mentors' time and expertise. Expressing gratitude was another essential element of networking etiquette that Sarah diligently practiced. She understood the importance of acknowledging her mentors' guidance and support. Whether through a sincere thank-you note, a heartfelt expression of appreciation during mentorship sessions, or by showcasing their impact in her achievements, Sarah made sure her mentors knew how grateful she was for their presence in her life. Sarah also recognized the value of reciprocity in networking etiquette. While mentors offered their guidance and support, she sought opportunities to give back and contribute to their success. Whether it was sharing relevant articles, connecting them with valuable resources, or providing support in their projects, Sarah made an effort to contribute to the lives and careers of her mentors. This reciprocity created a mutually beneficial dynamic and strengthened the foundation of their mentorship relationship.

Furthermore, Sarah made it a point to be reliable and respectful of her mentors' time. She understood that their schedules were often demanding, and she sought to be mindful of that. Sarah prepared diligently for their mentorship sessions, maximizing the time they spent together. She was punctual and respectful of agreed-upon meeting times, ensuring that she never took their availability for granted. In social settings, Sarah embodied professionalism and tact. She understood the importance of representing herself and her mentors in a positive light. Sarah dressed appropriately, maintained a friendly and approachable demeanour, and engaged in conversations with grace and humility. She recognized that networking events provided valuable opportunities to meet new professionals and expand her network, and she navigated these settings with confidence and authenticity. Through her commitment to networking etiquette, Sarah cultivated strong and lasting mentorship relationships. Her mentors appreciated her professionalism, respect and genuine interest in their guidance. As a result, they continued to invest in her growth and provided ongoing support throughout her journey.

As you onboard on your own mentorship relationships, remember the importance of networking etiquette. Actively listen, express gratitude, practice reciprocity, be reliable and embody professionalism. By honouring these principles, you will not only nurture meaningful connections but also strengthen the bond with your mentors, allowing them to guide you effectively on your path to success. Sarah's story underscores the significance of networking etiquette in mentorship relationships. Her commitment to practising good networking etiquette paved the way to deep, meaningful connections that transcended professional boundaries. Let her journey inspire you to embrace networking etiquette as an essential aspect of your mentorship journey, fostering relationships built on respect, gratitude and professionalism.

KEY TAKEAWAYS

CHAPTER 9: BUILDING A STRONGER NETWORK: HOW MENTORSHIP CAN HELP YOU EXPAND YOUR PROFESSIONAL AND PERSONAL CONNECTIONS

How Mentorship Can Help You Build a More Diverse Network

Mentorship is a transformative tool that can significantly expand both your professional and personal networks. By engaging in a mentor-mentee relationship, you gain access to diverse networks, receive guidance in relationship-building, benefit from recommendations and endorsements, and receive personalized introductions and referrals. Working closely with a mentor builds your confidence and credibility, while their long-term support and sponsorship contribute to network expansion over time.

Furthermore, mentorship can help you build a more diverse network, fostering inclusivity and expanding opportunities. Mentors bring different perspectives and experiences, introducing you to diverse professionals from various backgrounds, ethnicities, genders, and industries. This exposure enhances your understanding, empathy, and adaptability. Additionally, mentorship provides opportunities for cross-cultural learning, access to diverse communities and resources, and challenges biases while fostering inclusion. A diverse network brings forth a range of perspectives and opportunities for collaboration and growth.

To maximize the benefits of mentorship, seek out mentors who value diversity and inclusion actively engage in activities that promote diversity, and belong to diverse networks. By actively participating in a mentor-mentee relationship with a focus on diversity, you can expand your network, increase your cultural competence, and contribute to creating a more inclusive professional and personal sphere.

Strategies For Finding Mentors Outside of Your Immediate Network

1. Embrace curiosity and attend industry events to connect with experienced professionals.
2. Cultivate an abundance mindset, recognizing that mentors can be found in unexpected places.
3. Utilize technology to expand your reach and engage with professionals in online communities.
4. Seek out specialized communities and organizations aligned with your interest.
5. Harness the power of introductions by reaching out to your existing connections.
6. Stay motivated, persistent, and willing to step outside your comfort zone.

By following these strategies, you can find mentors outside your immediate network and experience transformative growth in your personal and professional connections.

Leveraging Mentorship to Build Relationships With Other Professionals In Your Field

Leveraging mentorship to build relationships with professionals in your field is a transformative strategy. By actively engaging with mentors' network, attending networking events, giving back and nurturing connections over time, you can create a vast and supportive network. These relationships provide collaborative opportunities, joint ventures, and increased influence within your industry. Remember, the relationships you cultivate are as important as mentorship itself. Embrace the power of mentorship to connect with like-minded individuals who will support and inspire you on your journey to success.

How Mentorship Can Help You Build Meaningful Personal Connections

Mentorship goes beyond professional growth; it helps build meaningful personal connections. Sarah's journey highlights how mentors cared for her overall well-being, offering advice in work-life balance and personal development. Mentorship opened doors to like-minded individuals who became lifelong friends. Sarah formed deep personal connections with individuals who shared her vision, becoming partners in community initiatives. The support from mentors positively impacted her personal relationships, making her more compassionate and confident. Mentorship enriched Sarah's life in immeasurable ways, extending beyond business matters.

Embrace mentorship for personal growth and deep connections that will enrich every aspect of your life.

The Importance of Networking Etiquette in Mentorship Relationships

Networking etiquette played a crucial role in Sarah's mentorship relationships, and she recognized its importance for nurturing and sustaining those connections. Sarah actively listened, asked thoughtful questions, and showed respect for her mentors' time and expertise. Expressing gratitude and practicing reciprocity were essential elements of her networking etiquette. She was reliable, punctual, and respectful of her mentors' schedules. In social settings, she embodied professionalism and engaged with grace and humility. Sarah's commitment to networking etiquette strengthened her mentorship relationships and garnered ongoing support. Remember the significance of networking etiquette in your own mentorship journey, as it fosters meaningful connections build on respect, gratitude, and professionalism.

CHAPTER 10: THE ROLE OF MENTORS IN LEADERSHIP DEVELOPMENT: HOW MENTORSHIP CAN HELP YOU BECOME A BETTER LEADER

> *"The delicate balance of mentoring someone is not creating them in your own image, but giving them the opportunity to create themselves"* – Steven Spielberg

We will explore the invaluable role of mentors in shaping exceptional leaders and how mentorship can elevate your leadership skills to new heights. Whether you are aspiring to lead a team, organization, or community, the guidance and support of a mentor can accelerate your growth and unleash your full leadership potential. Mentorship is a powerful relationship built on trust, guidance, and knowledge-sharing between an experienced mentor and a mentee seeking personal and professional growth. It is a transformative process that facilitates learning, skill development, and self-discovery. Mentos act as catalyst for growth, challenging mentees to step outside their comfort zones and encouraging them to explore new perspectives and possibilities. Mentorship fosters personal and professional development by nurturing qualities such as resilience, adaptability and emotional intelligence.

Throughout your leadership journey, mentors play a crucial role in providing guidance and inspiration. They share their experiences, impart wisdom, and inspire mentees to overcome obstacles and seize opportunities. By observing their mentors' behaviors and actions, mentees can develop a strong sense of purpose and a clear vision for their own leadership journey. Constructive feedback is the cornerstone of leadership development. Mentors provide honest and insightful feedback to mentees, highlighting their strengths and areas for improvement. This feedback loop self-reflection enables mentees to refine their leadership skills, address weaknesses, and continuously grow.

Mentors possess extensive networks and connections with their respective industries or communities. They can introduce mentees to influential individuals, opening doors to new opportunities and collaborations. By expanding their networks, mentees can gain valuable insights, access resources, and build relationships that enhance their leadership potential. The mentor-mentee relationship is built on trust and rapport. Mentors create a safe and confidential space for mentees to share their challenges, aspirations, and vulnerabilities. This trust enables open communication, fostering an environment conducive to learning and growth. Effective mentors recognize that each mentee has unique strengths, goals, and development areas. They tailor their guidance and support to meet the specific needs of the mentee, employing various mentoring approaches, such as coaching, advising and role modeling, to maximize the mentee's growth potential. The ultimate goal of mentorship is to empower mentees to become self-reliant leaders. Mentors gradually encourage mentees to think independently, make informed decisions, and take ownership of their leadership journeys. By fostering independence, mentors equip mentees with the confidence and capabilities to navigate future challenges successfully. Mentorship plays a vital role in leadership development. The guidance and support of a mentor can accelerate your growth, enhance your skills, and help you become a better leader. Throughout this chapter, we will delve deeper into the various aspects of mentorship and explore practical strategies to leverage the power of mentorship in your leadership journey.

How Mentorship Can Help You Develop Your Leadership Skills

In this section, we will explore how mentorship can significantly contribute to developing your leadership skills through the powerful and inspiring story of a team in an Indian organization. This team faced various challenges and lacked crucial leadership abilities. Despite initial resistance from a mentor due to past failures, their dedication and willingness to learn ultimately led to a transformative mentorship experience.

In Bangalore, India, at the headquarters of a IT company. Team Phoenix found themselves grappling with a series of setbacks. Led by Neha Kapoor, the team was responsible for developing innovative software solutions. However, their lack of effective leadership skills hindered their progress and impeded their ability to meet project deadlines. Recognizing the need for guidance, Neha and her team approached a highly accomplished and respected leader, Dr. Arjun Verma, in the hopes of receiving mentorship. However, due to the team's history of failure and underperformance, Dr. Verma initially hesitated to take them under his wing. He had witnessed numerous instances where teams failed to implement his guidance effectively. Determined to prove themselves, Neha and her team passionately expressed their desire to learn and improve. Their dedication, coupled with Dr. Verma's appreciation for their willingness to take responsibility for their past mistakes, led him to reconsider. Dr. Verma decided to mentor Team Phoenix, but with condition – they must be open-minded, committed, and willing to implement his teachings wholeheartedly.

Under Dr. Verma's mentorship, the team embarked on a transformative journey. They first identified their core challenges, including poor time management, lack of effective communication, and a fragmented decision-making process. Dr. Verma introduced powerful leadership concepts and techniques to address these issues. One significant challenge Team Phoenix faced was their inability to manage time effectively. Dr. Verma taught them strategies, such as prioritization, delegation and setting realistic deadlines. He emphasized the importance of effective planning and encouraged the team to adopt time management tools and techniques. Another critical area of improvement was communication. The team struggled to communicate ideas, requirements, and progress effectively. Dr. Verma introduced them to active listening techniques, assertive communication, and regular team meetings. He emphasized the significance of clear and concise communication to foster collaboration and avoid misunderstandings.

One of the most profound lessons the team learned from Dr. Verma was the power of accountability and taking ownership of their actions. He instilled in them a sense of responsibility and encouraged them to view failure as learning opportunities rather than setbacks. This shift in mindset allowed the team to embrace challenges with resilience and perseverance. Over time, Team Phoenix began to implement the strategies and lessons taught by Dr. Verma. Their time management improved, leading to efficient project execution and timely deliveries. Communication within the team became more open, transparent, and effective, resulting in better collaboration and increased productivity. The mentorship experience not only equipped Neha and her team with essential leadership skills but also instilled confidence and a renewed sense of purpose. Under Dr. Verma's guidance, Team Phoenix turned their setbacks into stepping stones for growth and success.

Finding Mentors Who Can Help You Develop Specific Leadership Qualities

After being mentored by Dr. Arjun Verma, a prominent leader who played a crucial role in Team Phoenix's leadership development, he retired from the organization, leaving the team facing new challenges. Neha Kapoor, the team leader, recognized that the team still lacked essential qualities such as vision, integrity, motivation, effective presentation, and influencing skills. Determined to address these gaps, Neha embarked on a search for a mentor who could help Team Phoenix reach their full potential.

Neha conducted extensive research to find the ideal mentor who possessed the necessary leadership qualities to transform her team. After careful consideration, she discovered a powerful transformational leader named Bhagyashree Guhe. Known for her exceptional skills in visioning, integrity, motivation, effective presentation, and influencing, Bhagyashree Guhe was the perfect match for Team Phoenix's needs. Neha reached out to Bhagyashree Guhe and shared the challenges faced by Team Phoenix. Impressed by Neha's dedication and the team's technical prowess, Bhagyashree Guhe agreed to mentor Team Phoenix and unlock their full leadership potential. Under Bhagyashree, guidance, Team Phoenix embarked on a transformative journey. Bhagyashree instilled a clear and inspiring vision within the team, helping them define their purpose and align their efforts towards a common goal. Through mentorship sessions, she guided Team Phoenix in crafting a vision statement that encapsulated their aspirations and set the direction for their work. Bhagyashree also emphasized the importance of integrity in leadership. She taught Team Phoenix the value of upholding ethical principles, taking responsibility for their actions and fostering a culture of trust and transparency within the team. Through open discussions and practical exercises, Team Phoenix learned to embody integrity in their work, interactions, and decision-making processes.

Motivation was another area where Team Phoenix sought improvement. Bhagyashree empowered the team by helping them discover their intrinsic motivators and aligning their work with their personal and professional goals. Through coaching and motivational techniques, she ignited a sense of passion and purpose with each team member, fueling their drive to excel. Effective presentation and influencing skills were vital for Team Phoenix's growth. Bhagyashree conducted intensive workshops and provided personalized coaching to enhance the team's ability to articulate their ideas persuasively and influence both their internal team and external stakeholders. She taught them techniques to captivate the audience, deliver compelling messages and build strong relationships base on trust and respect. As a result of Bhagyashree mentorship. Team Phoenix experienced a remarkable transformation. They developed a clear and inspiring vision that served as guiding force in their work. Upholding integrity became a fundamental aspect of their team culture, fostering trust and accountability. The team members became highly motivated driven by their intrinsic desire to succeed. Their presentation and influencing skills improved significantly enabling them to effectively communicate their ideas and forge strong connections with clients and colleagues. Under Bhagyashree mentorship, Team Phoenix evolved into potential leaders. Neha Kapoor witnessed her team's growth and development, as they embraced their newfound leadership qualities and applied them to drive positive change within the organization.

The mentorship provided by Bhagyashree proved instrumental in Team Phoenix's development. Neha Kapoor's determination to find the right mentor led her to Bhagyashree, whose expertise in visioning, integrity, motivation, effective presentation and influencing transformed the team. Team Phoenix's journey exemplifies the transformative power of finding mentors who can help individuals and teams develop specific leadership qualities, paving way for their success growth. As Team Phoenix continued their journey under the mentorship of Bhagyashree, they began to exhibit remarkable leadership qualities and potential. Neha Kapoor witnessed a newfound sense of purpose and direction within the team, as they embraced their roles as leaders and influencers. One significant challenge that Team Phoenix faced was their ability to effectively lead and influence their peers and clients. Bhagyashree recognized this obstacle and worked closely with the team to enhance their leadership and influencing skills. Through personalized coaching sessions, Bhagyashree guided Team Phoenix in developing strong interpersonal skills and emotional intelligence. They learned to empathize with others, understand different perspectives, and communicate their ideas in a persuasive and influential manner. The team practiced active listening, honed their negotiation abilities, and mastered the art of building rapport with both internal and external stakeholders.

Additionally, Bhagyashree emphasized the importance of empathy and servant leadership. She encouraged Team Phoenix to lead with compassion, understanding the needs of their team members and supporting their growth and development. This approach fostered a positive team culture, build trust, and motivated team members to go above and beyond. Team Phoenix also faced the challenge of effectively presenting their technical expertise and ideas to clients and stakeholders. Bhagyashree conducted intensive training sessions on effective presentation techniques, helping the team structure their messages, engage their audience, and convey their technical knowledge with clarity and confidence. Under Bhagyashree guidance, Team Phoenix practiced delivering presentations, received constructive feedback, and refined their public speaking skills. They learned to adapt their communication style to different audiences, presenting complex technical concepts in a way that was easily understood and appreciated. Moreover, Bhagyashree instilled a strong sense of purpose and commitment to excellence within Team Phoenix. She encouraged them to set high standards for themselves and consistently strive for personal and professional growth. The team members began to take ownership of their work, seeking opportunities to improve and innovate in their respective fields. As a result of Bhagyashree mentorship, Team Phocnix experienced a profound transformation. They not only developed specific leadership qualities but also cultivated a holistic approach to leadership encompassing vision, integrity, motivation, effective presentation and influencing skills. Neha Kapoor witnessed her team's growth and development as they became potential leaders within the organization. The impact of Bhagyashree mentorship extended beyond the team, as Team Phoenix began to inspire and influence others through their exceptional leadership qualities and technical expertise.

How To Create A Mentorship Plan For Leadership Development

After experiencing the transformative power of mentorship under the guidance of Bhagyashree Guhe, Team Phoenix recognized the importance of creating a structured mentorship plan to continue their leadership development journey. Neha Kapoor, as the team leader, took the initiative to design a comprehensive mentorship program that would empower her team members to further enhance their leadership skills and reach their full potential.

1. **Identify Individual Development Needs:** Neha conducted one-on-one discussions with each team member to understand their specific leadership development needs and aspirations. By identifying their strengths, weaknesses, and areas for improvement, she gained insights into the individual areas where mentorship would be most beneficial.

2. **Set Clear Goals and Objectives:** Neha worked closely with each team member to set clear and actionable goals for their leadership development. These goals were aligned with the team's overall

vision and strategic objectives. They encompassed areas such as visioning, integrity, motivation, effective presentation, and influencing skills, reflecting the specific leadership qualities Team Phoenix aimed to cultivate.

3. **Seek Out Suitable Mentors:** Neha recognized the importance of finding mentors who possessed the desired leadership qualities and had the experience and expertise to guide her team members effectively. She reached out to leaders within the organization, industry experts, and external networks to identify potential mentors who could provide valuable guidance and support.

4. **Formalize Mentorship Relationships:** Once suitable mentors were identified, Neha facilitated the formalization of mentorship relationships. She organized introductory meetings between mentors and mentees to establish rapport, set expectations, and define the scope and frequency of mentorship interactions. Clear guidelines were established to ensure a structured and meaningful mentorship experience.

5. **Create a Mentorship Action Plan:** Neha collaborated with each mentor and mentee pair to create a customized mentorship action plan. This plan outlined specific activities, milestones, and timeliness for achieving the mentees' leadership development goals. It included a variety of approaches, such as regular one-on-one meetings, skills-building

workshops, job shadowing opportunities, and feedback sessions.

6. **Encourage Continuous Learning and Reflection:** Neha fostered a culture of continuous learning and reflection within Team Phoenix. She encouraged mentees to proactively seek feedback, reflect on their progress, and leverage their mentor's guidance to drive personal and professional growth. Regular check-ins and performance discussions were conducted to assess the effectiveness of the mentorship plan and make any necessary adjustments.

7. **Measure and Celebrate Success:** Neha implemented mechanisms to measure the impact of the mentorship program on the team's leadership development. Key performance indicators, such as improved communication skills, enhanced decision-making abilities, and increased team cohesion, were tracked and evaluated. As milestones were achieved and goals were met, Neha celebrated the team's successes, reinforcing their commitment to continuous improvement.

By Creating a well-structured mentorship plan for leadership development, Neha Kapoor and Team Phoenix ensured that their journey towards becoming effective leaders continued on a path of growth and success. The mentorship program provided ongoing guidance, support, and accountability, allowing team members to develop the specific leadership qualities they lacked reflection,

Team Phoenix was able to cultivate strong leadership skills and make a positive impact within their organization and industry.

Creating a mentorship plan for leadership development is essential for unlocking the full potential of individuals and teams. By identifying development needs, setting clear goals, finding suitable mentors, and fostering a culture of continuous learning, organizations can nurture their talent and cultivate effective leaders who can drive sustainable success. The mentorship journey of Team Phoenix under the guidance of Bhagyashree serves as an inspiring example of how a well-designed mentorship plan can accelerate leadership development and transform a team's capabilities. Through the implementation of the mentorship plan, Team Phoenix was able to address their lack of vision, integrity, motivation, effective presentation and influencing skills.

Common Leadership Development Challenges And How To Overcome Them

Throughout the mentorship journey of Team Phoenix under the guidance of Bhagyashree Guhe, they encountered several common challenges that are often faced during leadership development. However, with Bhagyashree's expertise and the team's commitment, they were able to overcome these challenges and continue their growth as effective leaders. One of the initial challenges Bhagyashree faced was gaining the trust and buy-in of Team Phoenix. As a new mentor, she had to prove herself to a team that had previously experienced setbacks and lacked confidence in their leadership abilities. To address this challenge, Bhagyashree focused on building strong relationships with each team member, demonstrating her knowledge and experience, and showcasing her genuine commitment to their development. Bhagyashree was able to earn the team's trust and establish herself as a valuable mentor. Another common challenge was resistance to change and stepping out of comfort zones. Team Phoenix, despite their technical expertise, was hesitant to embrace new leadership approaches and take risks. Bhagyashree understood the importance of pushing boundaries and encouraged the team to explore innovative ideas, challenge their assumptions, and take calculated risks. She created a safe and supportive environment where team members felt comfortable experimenting and learning from failures. Through consistent encouragement and by leading by example, Bhagyashree helped Team Phoenix overcome their resistance to change and embrace new leadership practices.

Team Phoenix also faced the challenge of limited resources and time constraints. As busy professionals, they struggled to find dedicated time for leadership development activities amidst their demanding work schedules. Bhagyashree recognized the importance of prioritizing leadership development and worked closely with Neha Kapoor to allocate dedicated time slots for mentorship interactions, workshops, and skill-building activities. She emphasized the value of investing in their own growth and development, inspiring Team Phoenix to make leadership development a priority and find creative ways to optimize their time and resources. Additionally, Bhagyashree identified the challenge of maintaining consistent motivation and sustaining the momentum of leadership development efforts. Team Phoenix experienced moments of frustration and setbacks, and it was crucial to keep their motivation high throughout the journey. Bhagyashree employed various strategies to address this challenge. She celebrated small wins, acknowledged individual and team achievements, and regularly reminded the team of their long-term vision and the positive impact they were making. By fostering a supportive and encouraging environment, Bhagyashree ensured that Team Phoenix stayed motivated and committed to its leadership development goals.

Furthermore, Team Phoenix encountered the challenge of effectively applying the leadership skills they were learning in real-world situations. While they were acquiring new knowledge and skills, the team needed guidance on how to translate this learning into practical action and overcome specific obstacles they faced in their roles. Bhagyashree provided individualized coaching, offered real-time advice, and encouraged the team to reflect on their experiences and apply the learned skills in their day-to-day work. She facilitated opportunities for team members to practice and refine their leadership skills, ensuring they could confidently navigate challenges and make meaningful contributions. To overcome these common leadership development challenges, Bhagyashree employed a holistic approach that encompassed building trust, embracing change, managing resources effectively, sustaining motivation and bridging the gap between theory and practice. Her guidance and mentorship helped Team Phoenix develop the resilience, adaptability and perseverance necessary to overcome these challenges and emerge as strong and influential leaders within their organization.

The Role of Self-Reflection In Leadership Development

As Team Phoenix continued their leadership journey under the mentorship of Bhagyashree Guhe, they discovered the profound impact of self-reflection on their growth as leaders. Self-reflection is a powerful tool that allows individuals to gain insights into their strengths, weaknesses, values, and beliefs. It fosters self-awareness and provides a foundation for personal and professional development. In the context of leadership, self-reflection plays a pivotal role in identifying areas for improvement, aligning actions with values, and enhancing overall leadership effectiveness. Self-reflection enables leaders to gain a deeper understanding of their leadership style and its impact on others. It prompts them to evaluate their communication strategies, decision-making processes, and conflict resolution skills. By engaging in introspection, leaders can identify their strengths and leverage them effectively, as well as recognize areas where growth and development are needed. This self-awareness allows leaders to make conscious choices about their behaviour and approach, leading to more impactful and authentic leadership.

Team Phoenix, guided by Bhagyashree, engaged in regular self-reflection exercises to enhance their leadership capabilities. They dedicated time to introspect on their leadership journey, examining their actions, decision, and interactions with team members and stakeholders. This reflection helped them recognize patterns, biases, and areas where they could improve their leadership effectiveness. Self-reflection also plays a crucial role in aligning actions with values. It allows leaders to clarify their core values and ensure that their decisions and behaviors are in alignment with those values. When leaders have a clear understanding of their values, they can lead with integrity and authenticity, earning the trust and respect of their team members and stakeholders. Through self-reflection, Team Phoenix members gained a deeper understanding of their personal values and consciously aligned their leadership practices with those values.

Furthermore, self-reflection enables leaders to develop a growth mindset and embrace a continuous learning mindset. By reflecting on their experiences and outcomes, leaders can extract valuable lessons and insights that fuel their growth and development. They become open to feedback, receptive to new ideas, and willing to challenge their assumptions. This mindset of continuous learning and improvement is vital for leadership development, as it allows leaders to adapt to changing circumstances, embrace innovation, and drive organizational success.

In the case of Team Phoenix, self-reflection played a significant role in their transformation as leaders. It allowed them to learn from their experiences, both successes and failures and make adjustments to their leadership approaches. Through self-esteem, they identified areas where they needed to strengthen their skills, such as effective presentation and influencing teams and clients. This awareness prompted them to seek out opportunities for learning and growth, both within and outside the mentorship program. To foster a culture of self-reflection within Team Phoenix, Bhagyashree encouraged regular moments of introspection and provided guidance on effective self-reflection techniques. She emphasized the importance of carving out dedicated time for reflection and creating a safe and non-judgmental space for team members to share their insights and reflections.

Self-reflection is a powerful tool in leadership development. It enables leaders to gain self-awareness, align actions with values, and foster a growth mindset. Through self-reflection leaders can identify areas for improvement, recognize the impact of their leadership style, and continuously learn and adapt. The mentorship journey of Team Phoenix under the guidance of Bhagyashree Guhe exemplifies the role of self-reflection in their transformation as leaders. By engaging in regular self-reflection exercises, Team Phoenix members were able to enhance their self-awareness, align their actions with their values, and drive their continued growth and development as effective leaders.

KEY TAKEAWAYS

CHAPTER 10: THE ROLE OF MENTORS IN LEADERSHIP DEVELOPMENT: HOW MENTORSHIP CAN HELP YOU BECOME A BETTER LEADER

How Mentorship Can Help You Develop Your Leadership Skills

1. Mentorship is a transformative process that accelerates personal and professional growth, unleashing your full leadership potential.
2. Mentors provide guidance, inspiration, and constructive feedback to develop your leadership skills.
3. Mentors expand your network, open doors to new opportunities, and facilitate collaborations.
4. Trust and rapport are crucial in the mentor-mentee relationship, creating a safe space for learning and growth.
5. The goal of mentorship is to empower mentees to become self-reliant leaders who make informed decisions and navigate challenges successfully.

Finding Mentors Who Can Help You Develop Specific Leadership Qualities

1. Finding mentors who possess specific leadership qualities can significantly enhance your team's development and success.

2. A mentor can help you develop qualities such as vision, integrity, motivation, effective presentation, and influencing skills.
3. Through personalized coaching and guidance, mentors empower teams to embrace leadership roles and drive positive change.
4. Leadership development includes enhancing interpersonal skills, emotional intelligence, and the ability to influence and inspire others.
5. Effective presentation techniques, adaptability to different audiences, and a commitment to excellence are essential for leadership growth.

How To Create A Mentorship Plan For Leadership Development

1. Identify individual development needs: Understand each team member's strengths, weaknesses, and areas for improvement to tailor the mentorship plan effectively.
2. Set clear goals and objectives: Align leadership development goals with the team's vision and strategic objectives, focusing on specific qualities to cultivate.
3. Seek out suitable mentors: Find mentors with desired leadership qualities and expertise to guide team members effectively, both within and outside the organization.
4. Formalize mentorship relationships: Establish rapport, set expectations, and define the scope and frequency of mentorship interactions through introductory meetings.

5. Create a mentorship action plan: Collaborate to develop a customized plan with specific activities and milestones for achieving leadership development goals.
6. Encourage continuous learning and reflection: Foster a culture of seeking feedback, reflecting on progress, and leveraging mentor guidance for personal and professional growth.
7. Measure and celebrate success: Track key performance indicators and celebrate milestones.

By implementing a well-structured mentorship plan, organizations can unlock the full potential of their talent, cultivate effective leaders, and drive sustainable success.

Common Leadership Development Challenges and How to Overcome Them

1. Building trust: Earn trust by establishing strong relationships, highlighting expertise and demonstrating genuine commitment to the team's development.
2. Embracing change: Encourage stepping out of your comfort zone, exploring innovative ideas, and creating a safe environment for experimentation and learning from failures.
3. Managing resources effectively: Prioritize leadership development by allocating dedicated time and finding creative ways to optimize resources.

4. Sustaining motivation: Celebrate achievements, remind the team of the long-term vision, and foster a supportive environment to keep motivation high.
5. Bridging theory and practice: Provide individualized coaching, offer real-time advice, and create opportunities to apply learned skills in real-world situations.

By addressing these common challenges, leaders can overcome obstacles and foster growth and development within their teams, enabling them to become influential and successful leaders.

The Role of Self-Reflection in Leadership Development

1. Self-reflection fosters self-awareness: It allows leaders to understand their strengths, weaknesses, and leadership style, enabling them to make conscious choices and improve their effectiveness.
2. Alignment with values: Self-reflection helps leaders clarify their core values and ensure their actions and decisions align with those values, leading to authentic and integrity-driven leadership.
3. Growth mindset and continuous learning: By reflecting on experiences, leaders extract valuable lessons, embrace feedback, and remain open to new ideas, fostering a mindset of growth and adaptation.
4. Learning from successes and failures: Self-reflection enables leaders to learn from both successes and failures, identifying areas for improvement and seeking out opportunities for growth.

5. Cultivating a culture of self-reflection: Leaders can encourage regular moments of introspection, provide guidance on effective techniques, and create a safe space for sharing insights and reflections.

By incorporating self-reflection into their leadership development journey, leaders can enhance their self-awareness, align their actions with values, foster continuous learning, and drive personal and professional growth.

CHAPTER 11: THE IMPACT OF TECHNOLOGY ON MENTORSHIP: HOW TO LEVERAGE DIGITAL TOOLS AND PLATFORMS FOR EFFECTIVE MENTORSHIP

> *"Our role as business leaders is to understand the strategic implications of digital transformation, lead the way forward and implement our vision for the future of our businesses and their contribution to the communities we support through our activities"* Antonio Belo Santos

In the digital age, technology has revolutionized various aspects of our lives, including mentorship. It has opened new avenues for mentorship, enabling individuals to connect, learn, and grow regardless of geographical barriers. The impact of technology on mentorship is undeniable, providing opportunities for effective mentorship that were previously unimaginable. Let us delve into an excellent inspiring example that showcases the power of leveraging digital tools and platforms for mentorship.

Meet Kiran Kumar, a young professional with a burning desire to excel in his career. He recognized the importance of mentorship in his professional growth but faced a significant challenge – he struggled to find a mentor who possessed the specific industry knowledge and expertise he sought. Undeterred, Kiran turned to the power of technology and discover a game-changing mentorship platform called "Mentor Connect". Mentor Connect was an innovative digital platform that connected individuals seeking mentorship with experienced professionals across various industries. Kiran signed up for the platform and was amazed by the wealth of mentors available to him, all with diverse backgrounds are expertise. The platform allowed him to search for mentors based on specific criteria, such as industry, skills, and location, ensuring a tailored and relevant mentorship experience. Kiran's mentor, Deepika Khanna, was an accomplished leader in the technology sector with a passion for mentoring professionals. Despite being located in a different city, technology bridged the distance between Kiran and Deepika. They leveraged video conferencing tools to conduct regular mentorship sessions, allowing for face-to-face interactions, even from afar. This virtual mentorship relationship enabled them to establish a strong connection and facilitated in-depth discussions about career goals, challenges, and strategies for growth.

One of the significant advantages of leveraging digital tools and platforms for mentorship is the flexibility it offers. Kiran and Deepika could schedule mentorship sessions at mutually convenient times, accommodating their busy schedules. This flexibility eliminated the limitations of time zones and enabled them to engage in mentorship discussions without any constraints. Additionally, they utilized instant messaging and email communication to maintain regular contact, seek guidance, and exchange resources and materials. The Mentor Connect platform also had a vibrant community of mentees and mentors, fostering peer-to-peer learning and networking. Kiran actively participated in discussion forums, webinars, and online events organized by the platform, where he had the opportunity to learn from other mentors and mentees. This community-driven approach expanded his knowledge base, exposed him to diverse perspectives, and broadened his professional network. Moreover, technology facilitated the sharing of resources and material between Kiran and Deepika. They utilized cloud storage platforms to exchange documents, articles, and relevant industry insights. Kiran benefited immensely from Deepika's curated resources, which supplemented their mentorship discussions and provided him with valuable learning materials.

The impact of technology on Kiran's mentorship journey was profound. It provided him with access to a pool of mentors who were otherwise geographically inaccessible. The virtual mentorship relationship with Deepika empowered Kiran to gain industry-specific knowledge, refine his career trajectory, and develop the necessary skills and competencies for success. Kiran's experience exemplifies the transformative power of leveraging digital tools and platforms for mentorship. Technology has democratized mentorship, making it more accessible and inclusive. It breaks down barriers of location, time and resources, connecting individuals with mentors who possess the expertise and insights they seek. Virtual mentorship relationships enable mentees to tap into a global network of mentors, unlocking opportunities for growth, learning, and professional development.

As technology continues to advance, it is crucial for aspiring leaders and mentees to embrace the vast possibilities it offers. By leveraging digital tools and platforms, individuals can connect with mentors from different backgrounds and perspectives, fostering diversity and enriching their mentorship experience. The impact of technology on mentorship is a testament to its ability to catalyze personal and professional growth, transcending limitations, and propelling individuals towards leadership excellence. The impact of technology on mentorship is transformative. Platforms like Mentor Connect provide individuals with access to a diverse pool of mentors, regardless of geographical constraints. Through video conferencing, instant messaging, and online communities, virtual mentorship relationships thrive, fostering learning, networking, and professional development. By embracing digital tools and platforms, aspiring leaders can harness the power of technology to unlock their full potential and pave the way for a successful mentorship journey.

The Benefits of Using Technology For Mentorship

Kiran Kumar's mentorship journey with Deepika Khanna through the Mentor Connect platform highlights the numerous benefits of leveraging technology for mentorship. Let us delve deeper into these benefits and explore how technology has revolutionized the mentorship landscape.

1. **Enhanced Accessibility:** Technology has broken down geographical barriers, making mentorship accessible to individuals regardless of their location. Traditional mentorship often required physical proximity between mentors and mentees, limiting opportunities for connection and learning. However, with the advent of virtual mentorship platforms like Mentor Connect, individuals can now seek guidance from mentors who possess the specific expertise they are looking for, irrespective of their physical proximity. Kiran, residing in a different city from Deepika, was able to connect with her seamlessly through video conferencing and digital communication tools, expanding his access to a broader range of mentorship opportunities.

2. **Flexibility and Convenience:** One of the significant advantages of technology-enabled mentorship is the flexibility and convenience it offers for both mentors and mentees. In a traditional mentorship setting, scheduling mentorship sessions could be challenging due to time constraints and conflicting schedules. However, virtual mentorship allows for greater flexibility, as individuals can schedule mentorship discussions at mutually convenient times. Kiran and Deepika utilized digital tools to find suitable time slots for their sessions, ensuring regular and meaningful interactions. This flexibility eliminates the barriers of time zones and conflicting

commitments, making mentorship more accessible and convenient for both parties involved.

3. **Diverse Perspectives and Global Networks:** Technology-enabled mentorship platforms often have a diverse community of mentors and mentees from various backgrounds and industries. This diversity fosters an environment of peer learning and networking, where mentees can gain insights from mentors with diverse perspectives and experiences. Kiran, through the Mentor Connect platform, had the opportunity to engage in discussion forums, webinars, and online events, expanding his knowledge base and networking with professionals from around the globe. This exposure to diverse perspectives enriches the mentorship experience, allowing mentees to broaden their horizons and develop a more comprehensive understanding of leadership.

4. **Resource Sharing and Collaboration:** Technology facilitates seamless sharing of resources and collaboration between mentors and mentees. In traditional mentorship models, exchanging documents, articles, and industry insights could be cumbersome and time-consuming. However, digital platforms provide an efficient means of sharing resources. Kiran and Deepika utilized cloud storage platforms to exchange relevant documents, articles, and industry insights. This sharing of resources supplemented their mentorship discussions and provided Kiran with valuable learning materials. Additionally, digital platforms also allow mentors to

curate and share resources with their mentees, ensuring a comprehensive and enriching mentorship experience.

5. **Continuous Support and Communication:** Digital tools enable mentors and mentees to maintain regular and consistent communication, ensuring ongoing support throughout the mentorship journey. Instant messaging, email and virtual meetings provide mentees with the means to seek guidance, ask questions and receive timely feedback from their mentors. The ease of communication provided by technology ensures that mentorship relationships remain strong, and mentees receive continuous support in their development journey. This constant interaction enhances the mentee's learning experience and fosters and deeper connection between mentors and mentees.

6. **Cost-Effectiveness:** Leveraging technology for mentorship often proves to be cost-effective compared to traditional in-person mentorship models. Virtual mentorship eliminates the need for travel expenses, accommodation, and other associated costs. This makes mentorship more accessible to individuals who may have budgetary constraints or limited resources. Technology-enabled mentorship platforms offer affordable options for individuals seeking guidance and development, making it a viable option for mentees from diverse backgrounds.

The benefits of using technology for mentorship extend beyond the story of Kiran Kumar and Deepika Khanna.

It is evident that technology has transformed mentorship offering new possibilities and avenues for individuals seeking guidance and growth. By embracing digital tools and platforms, mentees can tap into a global network of mentors, access diverse perspectives, and benefit from the flexibility and convenience that virtual mentorship provides. Technology has opened doors to new mentorship opportunities, breaking down barriers and democratizing access to mentorship for individuals worldwide.

The benefits of using technology for mentorship are manifold. Enhanced accessibility, flexibility, diverse perspectives, resource sharing continuous support, and cost-effectiveness are just a few of the advantages that technology brings to mentorship. By embracing and leveraging digital tools and platforms, individuals can embark on a transformative mentorship journey that transcends boundaries and propels them towards leadership excellence. The journey of Kiran Kumar and Deepika Khanna exemplifies how technology-enabled mentorship can revolutionize the way individuals connect, learn and grow, ultimately unlocking their true potential as leaders.

Best Practices For Using Digital Communication Tools In Mentorship Relationships

In the dynamic and ever-evolving landscape of mentorship, the utilization of digital communication tools has emerged as a game-changer, open up endless possibilities for mentorship relationships to flourish. The digital age has paved the way for unprecedented connectivity, allowing mentors and mentees to transcend geographical boundaries and connect on a profound level. To harness the transformative potential of digital communication tools in mentorship relationships, it is essential to embrace certain best practices that empower mentors and mentees to maximize their growth and impact. These best practices delve deep into the core of effective digital mentorship inspiring a powerful and meaningful connection that propels both parties toward excellence.

1. **Embrace Interactive Video Conference:** One of the most impactful ways to foster a genuine and engaging mentorship relationship is through interactive video conferencing. Utilize platforms such as Zoom, Microsoft Teams or Google Meet to simulate face-to-face interactions, despite being physically distant. The power of nonverbal cues, facial expressions, and body language can be preserved through video conferencing, enabling mentors and mentees to establish a stronger connection, and the cultivation of a mentorship experience that transcends limitations.

2. **Leverage Visual Content Sharing:** Visual content is a potent tool for mentorship, allowing for the sharing of knowledge and experiences in a captivating and digestible format. In addition to traditional text-based resources, mentors can utilize digital platforms like Dropbox, Google Drive, or OneDrive to share videos, infographics, presentations, or visual case studies. These visual aids enhance comprehension,

engagement, and retention of information, enabling mentees to absorb and apply insights with greater ease. By incorporating visual content sharing into the mentorship journey, mentors can inspire mentees, ignite their creativity, and impart knowledge in a more compelling and memorable manner.

3. **Foster Collaboration Through Cloud-Based Platforms:** Collaboration lies at the heart of effective mentorship. By leveraging cloud-based collaboration platforms such as Notion, Microsoft OneNote, or Trello, mentors and mentees can engage in joint projects, brainstorm ideas, and co-create valuable resources. These platforms provided a centralized space for collaboration, enabling seamless sharing, editing, and organizing of content. Through collaborative initiatives, mentors can empower mentees to develop critical thinking, problem-solving, and teamwork skills, while nurturing an environment of mutual respect and learning.

4. **Cultivate Active Online Communities:** In the digital realm, mentorship extends beyond the individual relationship between a mentor and a mentee. Actively participate in online communities, forums, and social media groups dedicated to mentorship and professional development. Engaging in these vibrant communities opens doors to diverse perspectives, networking opportunities, and a wealth of knowledge. Mentees can gain insights from seasoned professionals, interact with like-minded individuals, and expand their horizons. Similarly,

mentors can contribute their expertise, establish their thought leadership, and forge connections with potential mentees or other mentors. Actively nurturing an online presence within these communities' nurtures personal growth, cultivates professional networks, and amplifies the impact of mentorship.

5. **Emphasize Timely and Responsive Communication:** The speed of digital communication brings with it the expectation of timely and responsive communication. Mentors and mentees should prioritize prompt responses to messages, emails, or collaboration requests. This demonstrates respect, commitment, and genuine investment in the mentorship relationship. Regularly communicate progress updates, seek clarification, and provide feedback in a timely manner. By maintaining a high level of responsiveness, mentors and mentees can nurture trust, maintain momentum and keep the mentorship journey dynamic and impactful.

6. **Embrace the Power of Asynchronous Communication:** Asynchronous communication, where communication occurs without the need for immediate responses, can be a valuable asset in digital mentorship. Embrace tools like email, instant messaging apps, or project management software to foster ongoing communication that accommodates differing schedules and time zones. Asynchronous communication allows for thoughtful reflection,

deeper analysis, and the crafting of more comprehensive responses. It encourages mentees to formulate well-thought-out questions, and mentors can provide considered guidance and feedback. By embracing asynchronous communication, mentors and mentees can find a harmonious balance between continuous communication, and mentors and mentees can find a harmonious balance between continuous support and individual reflection, fostering a mentorship experience that values both connection and introspection.

7. **Encourage Self-Directed Learning and Resource Exploration:** Digital mentorship provides mentees with unparalleled access to a vast array of resources, enabling self-directed learning and exploration. Encourage mentees to proactively seek out industry-related articles, research papers, webinars, podcasts, or online courses. Facilitate access to digital libraries, online platforms, and subscription-based resources. By empowering mentees to take ownership of their learning journey, mentors can inspire them to become lifelong learners and equip them with the skills needed for continuous growth and adaptability in an ever-changing world.

8. **Cultivate a Culture of Feedback and Reflection:** Feedback is an essential ingredient for growth and development. Encourage open and honest feedback exchanges between mentors and mentees using digital tools like Google Forms, and Microsoft, Forms, or feedback platforms such as Slack or TINYpulse.

Regularly reflect on the mentorship journey, evaluating progress, discussing challenges, and setting new goals. Foster a safe and supportive environment that encourages mentees to seek feedback and mentors to provide constructive guidance. By cultivating a culture of feedback and reflection, mentorship relationships can continuously evolve, adapt, and inspire transformative growth.

Kiran Kumar and Deepika Khanna's mentorship journey exemplified the effective implementation of best practices for using digital communication tools in their relationship. As they embarked on their mentorship journey, they wholeheartedly embraced these practices, harnessing the power of technology to enhance their connection, learning, and growth. To foster meaningful and engaging interactions, Kiran and Deepika utilized interactive video conferencing as their primary mode of communication. They scheduled regular mentorship sessions via video conferencing platforms such as Zoom. This allowed them to simulate face-to-face interactions, despite being physically distant. Through video conferencing, they could observe each other's facial expressions and body language, creating a more profound and authentic connection. It enabled them to build trust, establish rapport, and engage in open and transparent discussions about Kiran's career goals, challenges, and strategies for growth. Additionally, Kiran and Deepika leveraged visual content sharing as a powerful tool in their mentorship journey. They utilized digital platforms like Dropbox and Google Drive to exchange videos, presentations, and visual case studies. Deepika shared industry-specific videos and presentations that enriched Kiran's learning experience. This visual content not only enhanced Kiran's comprehension but also sparked his creativity and inspired him to explore different perspectives within his field. The incorporation of visual content sharing deepened their mentorship conversations and facilitated a more immersive learning experience for Kiran.

Collaboration was another key aspect of Kiran and Deepika's mentorship relationship, enabled by cloud-based platforms. They utilized platforms such as Notion and Trello to collaborate on joint projects, brainstorm ideas, and co-create valuable resources. These platforms provided a centralized space for them to share ideas, documents, and feedback, fostering a sense of shared ownership and mutual growth. By engaging in collaborative initiatives, Kiran developed critical thinking, problem-solving, and teamwork skills, all while benefiting from Deepika's expertise and guidance. Actively participating in online communities dedicated to mentorship and professional development was another best practice embraced by Kiran and Deepika. Kiran engaged in online forums, webinars, and social media groups organized by the Mentor Connect platform and other industry-specific communities. He had the opportunity to learn from seasoned professionals, interact with like-minded individuals, and broaden his professional network. Deepika encouraged Kiran to share his insights and experiences within these communities, allowing him to establish his thought leadership and gain recognition among his peers. Their active involvement in online communities nurtured their growth, expanded their perspectives, and reinforced the power of mentorship in a broader context. Timely and responsive communication was a priority for Kiran and Deepika. They understood the significance of maintaining regular contact and promptly addressing each other's inquiries. They utilized instant messaging apps and email to stay connected between mentorship sessions, seeking guidance and sharing updates. This consistent communication ensured that Kiran received ongoing support and feedback from Deepika, keeping their mentorship journey dynamic and impactful.

Kiran and Deepika also recognized the value of asynchronous communication in their mentorship relationship. They understood that everyone's schedule and availability may vary. Therefore, they leveraged digital tools to engage in thoughtful and comprehensive exchanges that did not require immediate responses. By embracing asynchronous communication, Kiran had the time to reflect on Deepika's guidance, formulate well-thought-out questions, and seek advice based on his own pace and availability. This allowed for deeper introspection and fostered a more intentional mentorship experience. Throughout their mentorship journey, Kiran and Deepika emphasized the importance of feedback and reflection. They regularly evaluated their progress, discussed challenges, and set new goals. Utilizing digital feedback platforms, they exchanged constructive feedback that guided Kiran's growth and development. This culture of feedback and reflection created a safe and supportive environment where Kiran felt comfortable seeking guidance and Deepika could provide meaningful insights. The constant feedback and reflection ensured that their mentorship relationship remained impactful, adaptive, and transformative. As Kiran Kumar and Deepika Khanna concluded their mentorship journey, they marveled at the power of digital communication tools in elevating their mentorship experience. Through interactive video conferencing, visual content sharing, collaboration on cloud-based platforms, active participation in online communities, timely and responsive communication, embracing asynchronous communication, and fostering a culture of feedback and reflection, they harnessed the full potential of technology to foster a mentorship journey that transcended boundaries.

Their mentorship journey serves as an inspiring example of how the best practices for using digital communication tools in mentorship relationships can profoundly impact the growth and development of aspiring professionals. Kiran's knowledge, skills, and confidence flourished under Deepika's guidance, and the power of digital tools amplified their connection and learning. Kiran Kumar and Deepika Khanna's mentorship journey exemplified the transformative potential of leveraging digital communication tools in mentorship relationships. Their experience reinforces the importance of embracing these best practices and harnessing the power of technology to unlock new avenues of growth, knowledge, and professional development. Their story stands as a testament to the remarkable impact of digital mentorship in the modern era, inspiring mentees and mentors alike to embark on their own transformative mentorship journeys.

How To Use Online Platforms To Find And Connect With Mentors

Finding the right mentor who aligns with your career goals and possesses the expertise you seek can be a transformative experience. Online platforms have made this process more accessible, enabling individuals like Kiran Kumar to connect with mentors across various industries. Here's a step-by-step guide on how to use online platforms effectively to find and connect with mentors. First and foremost, it is important to research and choose the right platform that suits your mentorship needs. Start by exploring mentorship platforms that cater to your industry or area of interest. Look for platforms with a diverse pool of experienced mentors and positive reviews from past users. Take into consideration factors such as the platform's reputation, ease of use, available features, and success stories of mentees who have found valuable guidance through the platform. Once you have identified a suitable platform, sign up and create a compelling profile that highlights your aspirations and what you are looking for in a mentor. Next, it is crucial to identify your goals and criteria for mentorship. Take the time to reflect on your career aspirations and the specific expertise you require to achieve them. Define the skills industry, and experiences you want your mentor to possess. Having clear criteria will help you filter potential mentors effectively and find the right match for your development needs.

Leverage the advanced search filters provided by the platform to narrow down your mentor options. Filter mentors based on their industry, location, years of experience, and areas of expertise. This ensures that you find mentors who closely align with your requirements, increasing the likelihood of a successful mentorship relationship. Once you have identified potential mentors, review their profiles thoroughly. Pay attention to their background, accomplishments, and mentoring style. Look for mentors whose values and approach resonate with you. Some platforms may also provide mentee reviews or ratings for mentors, which can offer valuable insights into the mentor's effectiveness and impact. Take the time to read this review and gauge the mentor's compatibility with your needs.

With potential mentors identified, it is time to initiate the connection. Craft a personalized message expressing your interest in their mentorship. Highlight why you believe they are the right fit for your goals and what you hope to gain from the mentorship. A genuine and well-crafted message increases the chances of positive response. When reaching out to potential mentors, maintain a professional tone and show respect for their time and expertise. Avoid using overly casual language or making demands. Instead, demonstrate your eagerness to learn and your appreciation for their willingness to share their knowledge Be concise and focused in your communication, clearly articulating your goals and expectations. If a mentor responds positively to your connection request, promptly follow up and express gratitude for their willingness to engage in mentorship.

Discuss potential meeting times and preferences for communication to establish a schedule that works for both parties. Be flexible and accommodating in finding a suitable time for mentorship sessions. Consider connecting with multiple potential mentors initially, as not every connection may lead to a fruitful mentorship relationship. But exploring multiple options increases the likelihood of finding a mentor with whom you have a strong connection and shared vision. However, it is essential to be respectful of the mentor's time and not overextend your commitment to multiple mentors simultaneously. Once you have established a mentorship relation, show commitment and dedication to the process. Be prepared for mentorship session, ask thoughtful questions, and actively engage in discussion. Implement the guidance and feedback received from your mentor and demonstrate progress in your professional growth. Treat each mentorship session as a valuable opportunity for learning and development.

To express gratitude for your mentor's time, support, and guidance. A simple thank you note, or a token of appreciation can go a long way in strengthening your mentorship bond and demonstrating your respect for their investment in you development. Lastly leverage the resources and networking opportunities provided by the online platform. Engage with the platform's community, participate in discussion forums, attend webinars, and explore additional learning materials and resources. Actively seek out opportunities to connect with other mentors and mentees to expand your network and gain insights from diverse perspectives. Online platforms have revolutionized the way individuals find and connect with mentors. By following these steps, aspiring mentees like Kiran Kumar can navigate the online mentorship landscape effectively. With careful research, thoughtful communication, and a commitment to growth, individuals can find mentors who align with their goals and unlock the transformative power of mentorship, Embrace the opportunities provided by online platforms, and embark on a mentorship journey that propels you toward success.

Common Challenges In Using Technology For Mentorship And How To Overcome Them

1. **Limited Personal Connection:** One of the main challenges in using technology for mentorship is the potential loss of personal connection. Traditional mentoring often relies on fact-to-face interactions, body language, and non-verbal cues to build a strong bond between mentors and mentees. However, technology-mediated mentorship can lack the same level of intimacy and immediacy. To overcome this challenge, it is crucial to leverage technology tools that facilitate real-time communication and foster a sense of connection. Video conferencing platforms such as Zoom or Microsoft Team can be utilized for virtual face-to-face interactions. Mentors and mentees should also make an effort to establish a regular meeting schedule and engage in active listening to bridge the gap created by the virtual medium.

2. **Time Zone and Availability Constraints:** Mentorship relationships often span across different time zones, making it challenging to find mutually convenient meeting times. Scheduling conflicts and availability constraints can hinder regular and effective communication. To address this challenge, mentors and mentees can use online scheduling tools to find suitable meeting times that accommodate their respective time zones. Flexibility and understanding are key in navigating time zone differences. Mentors and mentees can also explore asynchronous communication methods, such as email or instant messaging, to stay connected and address questions or concerns even when immediate real-time interactions are not possible.

3. **Digital Divide and Access:** The digital divide refers to the gap in access to technology and digital resources. Not all individuals have equal access to computers, high-speed internet, or the necessary software and devices for effective technology-mediated mentorship. To overcome this challenge, mentos and organizations should strive to provide equal access to technological resources for all participants. This can include offering loaner laptops, internet connectivity assistance or identifying local community resources that provide access to technology. Mentors can also consider using low-bandwidth communication methods, such as text-based messaging or phone calls, to ensure inclusivity and reach individuals with limited access to advanced technology. While technology offers numerous

benefits for mentorship, it also presents challenges that need to be addressed for effective implementation. By recognizing and proactively tackling these challenges, mentors and mentees can leverage technology to create meaningful and impactful mentorship relationships, irrespective of physical distances.

The Future Of Technology And Mentorship

As technology continues to evolve at a rapid pace, its impact on mentorship is poised to grow exponentially. The future of technology and mentorship holds tremendous potential for transforming the way mentorship is conducted and experienced. Here are some exciting possibilities.

1. **Virtual Reality (VR) and Augmented Reality (AR):** Virtual reality and augmented reality technologies have the potential to revolutionize the mentorship experience. Imagine mentees being immersed in virtual environments where they can practice real-life scenarios or receive hands-on training. VR and AR can create highly interactive and engaging mentorship experiences that bridge the gap between theory and practice. For example, a mentee interested in learning a technical skill could don a VR headset and interact with a virtual mentor who guides them through a simulated training program. This immersive experience can enhance learning outcomes and provide mentees with valuable practical knowledge.

2. **Artificial Intelligence (AI) and Machine Learning:** AI and machine learning technologies can significantly enhance the mentorship process by providing personalized guidance and insights. AI-powered mentorship platforms can analyze vast amounts of data to offer tailored recommendations and suggestions for mentees based on their individual needs and goals. AI chatbots can act as virtual mentors, providing instant feedback, answering questions, and offering guidance 24/7. These intelligent systems can simulate human-like conversations, adapting their responses based on the mentee's progress and specific areas of focus. AI algorithms can also identify patterns in mentoring relationships, highlighting potential areas for improvement, and suggesting strategies for enhancing the mentorship experience.

3. **Collaborative Platforms and Gamification:** Future mentorship platforms are likely to incorporate collaborative tools and gamification elements to foster engagement and motivation. Collaborative platforms can facilitate mentor-mentee interactions, enabling seamless communication and resource sharing. Mentees can collaborate with peers, mentors, and experts from various fields, expanding their networks and gaining diverse perspectives. Gamification techniques, such as badges, leaderboards, and rewards, can make the mentorship journey more enjoyable and encourage mentees to actively participate and achieve milestones. Virtual challenges, simulations, and gamified learning

modules can add an element of fun and excitement, promoting active learning and skill development.

4. **Data Analytics and Insights:** The integration of data analytics in mentorship programs can provide valuable insights into the effectiveness of mentoring relationships and identify areas for improvement. By analyzing data on mentee progress, engagement levels, and feedback, organizations can refine their mentorship strategies and optimize the outcomes. Data-driven insights can help mentors identify mentees who may require additional support, track their growth trajectory, and tailor their guidance accordingly. Advanced analytics can also identify patterns of success and highlight best practices, enabling organizations to replicate successful mentoring models.

5. **Cross-Cultural Mentorship:** Technology has the power to bridge geographical and cultural barriers, opening opportunities for cross-cultural mentorship. Virtual mentorship programs can connect mentees with mentors from diverse backgrounds and across different parts of the world, facilitating cross-cultural learning. Cross-cultural mentorship can provide mentees with global perspectives, expose them to different cultural norms, and foster inclusivity. It can broaden mentees' horizons, enhance their intercultural competencies, and prepare them for the increasingly globalized and interconnected world.

KEY TAKEAWAYS

CHAPTER 11: THE IMPACT OF TECHNOLOGY ON MENTORSHIP: HOW TO LEVERAGE DIGITAL TOOLS AND PLATFORMS FOR EFFECTIVE MENTORSHIP

The Benefits Of Using Technology For Mentorship

1. Technology has revolutionized mentorship by breaking down geographical barriers and providing access to mentors from diverse backgrounds and expertise.

2. Virtual mentorship platforms offer flexibility and convenience, allowing mentees to schedule sessions at mutually convenient times, irrespective of time zones.

3. Technology fosters diversity and global networking, exposing mentees to diverse perspectives and expanding their professional networks.

4. Digital tools enable seamless sharing of resources and collaboration between mentors and mentees, enhancing the mentorship experience.

5. Continuous communication through instant messaging and virtual meetings ensures ongoing support and guidance throughout the mentorship journey.

6. Technology-enabled mentorship is cost-effective, eliminating the need for travel expenses and making mentorship accessible to individuals with limited resources.

By embracing technology and leveraging digital tools and platforms, individuals can tap into the transformative power of mentorship, connecting with mentors worldwide, gaining diverse perspectives, and unlocking their true leadership potential.

Best Practices For Using Digital Communication Tools In Mentorship Relationships

1. Embrace Interactive video conference: Utilize platforms like Zoom, Microsoft Teams, or Google Meet to simulate face-to-face interactions and establish a stronger connection despite physical distance.

2. Leverage visual content sharing: Share videos, infographics, presentations, or visual case studies using platforms such as Dropbox, Google Drive, or One Drive to enhance.

3. Foster Collaboration through cloud-based platforms: Use platforms like Notion, Microsoft, One Note, or Trello to engage in joint projects, brainstorm ideas, and co-create valuable resources.

4. Cultivate active online communities: Participate in online communities, forums and social media groups

dedicated to mentorship and professional development for diverse perspectives, networking, and knowledge sharing.

5. Emphasize timely and responsive communication: Prioritize prompt responses to messages, emails, or collaboration requests to demonstrate commitment and maintain momentum in the mentorship relationship.

6. Embrace the power of asynchronous communication: Utilize tools like email, instant messaging apps, or project management software to accommodate differing schedules and time zones, allowing for thoughtful reflection and comprehensive responses.

7. Encourage self-directed learning and resource exploration; Empower mentees to proactively seek out industry-related resources such as articles, webinars, or online courses to become lifelong learners.

8. Cultivate a culture of feedback and reflection: Foster open and honest feedback exchanges using digital tools like Google Forms, and Microsoft Forms, or feedback platforms such as Slack or TINYpulse, and regularly reflect on the mentorship journey to evaluate progress and set new goals.

Kiran and Deepika's mentorship journey showcased how they applied these best practices. They utilized interactive video conferencing for meaningful interactions, visual content sharing for immersive learning, cloud-based platforms for collaboration, and engaged online communities for networking and learning. They also prioritized timely communication, embraced asynchronous communication, encouraging self-directed learning, and fostered a culture of feedback and reflection.

Overall, Kiran and Deepika's mentorship journey demonstrates the transformative impact of digital communication tools in mentorship relationships and serves as an inspiring example for others to leverage technology for growth, knowledge sharing, and professional development.

How To Use Online Platforms To Find and Connect With Mentors

Using online platforms to find and connect with mentos can be a transformative experience in your professional growth. Here is a step-by-step guide on how to effectively leverage online platforms for mentorship:

1. Research and Choose the Right Platform: Explore mentorship platforms that cater to your industry or area of interest. Consider factors such as reputation, user reviews, ease of use, and success stories of mentees. Create a compelling profile that highlights your aspirations and what you're looking for in a mentor.

2. Identify Your Goals and Criteria: Reflect on your career aspirations and the specific expertise you seek in a mentor. Define the skills, industry and experiences you want your mentor to possess.

3. Use Advanced Search Filters: Utilize the platform's advanced search filters to narrow down potential mentors based on industry, location, experience and expertise.

4. Review Mentor Profiles: Thoroughly review potential mentors' profiles, considering their background, accomplishments, and mentoring style. Look for mentors whose values and approach align with your needs. Read mentee reviews or rating if available.

5. Initiate the Connection: Craft a personalized message expressing your interest in mentorship. Clearly articulate why you believe they are the right fit and what you hope to gain. Maintain a professional tone and show respect for their time and expertise.

6. Establish the Connection: If a mentor responds positively, follow up promptly and discuss potential meeting times and preferences for communication. Be flexible and accommodating in finding a suitable schedule. Connect with multiple potential mentors initially to explore options but respect the mentor's time.

7. Show Commitment and Dedication: Prepare for mentorship sessions, ask thoughtful questions, and

actively engage in discussions. Implement guidance and feedback received, demonstrating progress in your professional growth.

8. Express Gratitude: Thank you mentor for their time, support, and guidance. A simple note of appreciation strengthens the mentorship bond and shows respect for their investment in your development.

9. Leverage Platform Resources and Networking: Engage with the platform's community, participate in forums, attend webinars, and explore additional learning materials. Connect with other mentors and mentees to expand your network and gain diverse perspectives.

By following these steps, you can effectively leverage online platforms to find and connect with mentors who align with your goals and unlock the transformative power of mentorship. Embrace the opportunities provided by online platforms and embark on a mentorship journey that propels you toward success.

Common Challenges In Using Technology For Mentorship And How To Overcome Them

Solutions: Real-time communication tools, regular meeting schedules, active listening. Online scheduling tools, asynchronous communication. Equal access to technology, support for limited access. Overcome challenges for impactful mentorship.

The Future Of Technology And Mentorship

1. Virtual Reality (VR) and Augmented Reality (AR): Immersive experiences for hand-on training and simulations.

2. Artificial Intelligence (AI) and Machine Learning: Personalized guidance and 24/7 virtual mentors.

3. Collaborative Platforms and Gamification; Seamless Communication, resource sharing, and engaging learning experiences

4. Data Analytics and Insights: Refining mentorship strategies based on data-driven insights.

5. Cross-Cultural Mentorship: Connecting mentees with mentors from diverse backgrounds for global perspectives and inclusivity.

Technology will revolutionize mentorship, enhancing engagement, personalization, and global connectivity.

CHAPTER 12: MENTORSHIP AND DIVERSITY: HOW TO FIND AND ENGAGE MENTORS FROM DIFFERENT BACKGROUNDS AND PERSPECTIVES

"True mentorship thrives in the embrace of diversity, where mentors and mentees from different backgrounds and perspectives come together to ignite inspiration, broaden horizons, and unlock the full potential within each other." – Maya Angelou

In the year of 2017, September 2017 in the bustling city of Mumbai, there resided a determined young man named Sreekanth. He had embarked on a transformative journey, seeking a mentor who could guide him towards success and empower him to reach his full potential. Sreekanth believed that a mentor with a diverse background and the ability to connect with individuals from different walks of life would be instrumental in his growth.

Driven by a deep desire to find the perfect mentor, Sreekanth embarked on an arduous search. He encountered numerous mentors along the way, but none seemed to meet his expectations. However, Sreekanth remained undeterred, knowing that the right mentor was waiting to be found. After months of relentless pursuit, Sreekanth's perseverance led him to Bhagyashree, a renowned business leader with a diverse background and a reputation for empowering and inspiring individuals from various perspectives. Sreekanth felt an instant connection with Bhagyashree's record and reputation and approached her with a request for mentorship.

At first, Bhagyashree was hesitant to take on another mentee. She had been approached by many individuals in the past but had found it challenging to establish a deep connection. However, she sensed something different in Sreekanth's approach and decided to put his dedication and consistency to the test. Bhagyashree gave Sreekanth a series of small tasks to complete within specific deadlines. The first task required meticulous attention to detail and thorough research. Sreekanth approached it with unwavering focus and dedication, pouring his heart and soul into ensuring its successful completion. To Bhagyashree's delight, he not only met the deadline but exceeded her expectations. Encouraged by Sreekanth's commitment, Bhagyashree presented him with a more challenging second task. This time, the deadline was tighter, and the task itself demanded creative problem-solving skills. Sreekanth faced obstacles and struggled along the way, but his determination propelled him forward. With sheer perseverance and a last-minute push, he managed to complete the task just moments before the deadline. Bhagyashree recognized the strength of Sreekanth's character. His consistent dedication and unwavering focus despite facing challenges had left a lasting impression on her. However, she wanted to test his resilience further before committing to mentorship.

For the thirst task, Bhagyashree designed a completely different challenge. It required Sreekanth to think outside the box and find innovative solutions. Sreekanth approached the task with the same level of dedication and determination as before. However, despite his best efforts, he found himself unable to achieve the desired outcome. Sreekanth felt disheartened, thinking his failure would end his chances of being mentored by Bhagyashree. Contrary to his expectations. Bhagyashree's perspective shifted. She saw in Sreekanth a rare quality-a willingness to push boundaries and embrace challenges head-on, even in the face of failure. This resilience convinced her that Sreekanth was indeed the mentee she had been searching for. She acknowledged his focus, dedication, and consistent efforts, and agreed to take him under her wing as a mentor. This is the Sreekanth's transformative journey, a testament to the power of mentorship and the significance of diversity in finding the right mentor. Through his unwavering dedication and commitment, Sreekanth not only proved his worth to Bhagyashree but also inspired others with his persistence and resilience.

Their mentorship journey had just begun, and Sreekanth knew that under Bhagyashree's guidance, he would continue to grow and make a significant impact. Together, they would navigate the complexities of diversity and use their unique perspectives to create a positive change in the world. In the realm of mentorship and diversity, Sreekanth's story serves as a reminder that finding and engaging mentors from different backgrounds and perspectives can open doors to incredible growth and transformation. Through dedication, focus, and a willingness to embrace challenges, individuals can forge powerful connections and embark on a journey of lifelong learning and empowerment.

The Importance Of Diversity In Mentorship Relationships

Sreekanth's transformational journey with Bhagyashree as his mentor not only emphasized the importance of diversity but also addressed the pain points, problems, and provided solutions within their mentorship relationship. Prior to finding Bhagyashree, Sreekanth had struggled to find a mentor who truly understood his unique challenges and aspirations. He had faced moments of self-doubt and uncertainty, unsure of how to navigate the obstacles on his path to success. However, his encounter with Bhagyashree marked a turning point. Through open and honest communication, Sreekanth shared his pain points and challenges with Bhagyashree. He spoke about the difficulties he had encountered while searching for a mentor who could understand and relate to his specific background and aspirations. Bhagyashree, as an empathetic and inclusive mentor, listened intently and acknowledged the significance of his struggles.

Through open and honest communication, Sreekanth shared his pain points and challenges with Bhagyashree.

He spoke about the difficulties he had encountered while searching for a mentor who could understand and relate to his specific background and aspirations. Bhagyashree, as an empathetic and inclusive mentor, listened intently and acknowledged the significance of his struggles. Together, they explored potential solutions to address these pain points. Bhagyashree encouraged Sreekanth to embrace diversity in mentorship, explaining that finding mentors from different backgrounds and perspectives could provide him with the guidance and insights he needed. They discussed strategies to identify mentors who could bridge the gaps in Sreekanth's knowledge and experiences, ensuring that his journey towards success would be more supported and well-rounded. Moreover, Bhagyashree understood the importance of representation and the impact it could have on Sreekanth's journey. She recognized that the lack of role models who shared his background could be discouraging and isolating. To address this, Bhagyashree encouraged Sreekanth to seek out community organizations, networks, and professional groups that focused on empowering individuals from similar backgrounds. This allowed Sreekanth to connect with like-minded individuals, share experiences, and find mentors who had successfully overcome similar obstacles.

Through their mentorship, Sreekanth gained the tools and confidence to address his pain points. He learned to articulate his challenges, seek support from diverse mentors, and navigate the complexities of his unique journey Bhagyashree's guidance and insights helped him overcome obstacles and empowered him to develop resilience in the face of adversity. Their mentorship relationship became a safe space for Sreekanth to share his struggles, seek guidance, and find practical solutions. Together, they celebrated his victories, learned from his setbacks, and continuously worked towards creating a supportive environment that encouraged growth and diversity. Sreekanth's journey exemplifies the transformative power of addressing pain points and finding effective solutions within a mentorship relationship. Through the recognition of his unique challenges and the implementation of tailored strategies, Sreekanth was able to overcome barriers and pave his way towards success. In the real of mentorship and diversity, Sreekanth's journey serves as a powerful reminder that acknowledging pain points, seeking out diverse mentors, and fostering an inclusive environment are integral to empowering individuals on their path to success. Bhagyashree's mentorship not only provided guidance but also addressed the specific challenges faced by Sreekanth, ensuring his personal growth and inspiring others to find mentors who can help them navigate their own pain points and achieve their full potential.

How To Find Mentors From Different Backgrounds And Perspectives

Within Sreekanth's transformative journey, a specific problem arose that encapsulated the need for mentors from different backgrounds and perspectives. This problem centered around the lack of representation and support for individuals from marginalized communities within his chosen industry. Sreekanth experienced a deep pain, witnessing the underrepresentation of individuals from diverse backgrounds in leadership positions. He recognized the systemic barriers that hindered their progress and felt compelled to make a difference. Sreekanth understood that the solution to this problem lay in finding mentors who could guide him through the challenges and help him break these barriers. To address this pain point, Bhagyashree and Sreekanth developed a specific solution. They focused on identifying mentors from marginalized communities who had successfully navigated the industry and achieved leadership positions. This approach allowed Sreekanth to connect with mentor who could not only offer diverse perspectives but also understood firsthand the obstacles he faced.

Bhagyashree guided Sreekanth in seeking out mentors who had shattered glass ceiling and emerged as trailblazers within their fields. They researched industry pioneers, attended panel discussions and conferences focused on diversity and inclusion, and actively sought out mentors from underrepresented backgrounds. This deliberate search allowed Sreekanth to find mentors who had faced similar challenges and could offer invaluable insights into overcoming them. The mentors Sreekanth connected with shared their experiences of breaking through barriers, navigating bias, and establishing themselves as influential leaders. They provided guidance on building networks, developing personal branding, and leveraging strengths unique to their cultural backgrounds. Through their mentorship, Sreekanth gained not only industry-specific knowledge but also a profound sense of empowerment and resilience. The solution to the Sreekanth's pain point became twofold: first, finding mentors from diverse backgrounds, and second, specifically seeking mentors who had overcome the barriers that individuals from marginalized communities often face. By connecting with mentors who could empathize with his struggles, Sreekanth received the guidance and support he needed to navigate the industry and actively work towards breaking down systemic barriers. This approach proved to be transformative for Sreekanth. Armed with the knowledge, support and inspiration from his mentors, he became a catalyst for change within his industry. Sreekanth established initiatives that promoted diversity, encouraged representation, and provided mentorship opportunities for individuals from marginalized backgrounds. His efforts aimed to create a more inclusive and equitable industry landscape for future generations.

Sreekanth's journey not only highlights the power of seeking mentors from different backgrounds and perspectives but also emphasized the significance of addressing specific pain point. By actively pursuing mentors who had triumphed over the barriers faced by marginalized communities, Sreekanth found a solution that empowered him to bring about systemic change. This experience serves as an inspiration for individuals who strive to make a meaningful impact. It demonstrates that by finding mentors who understand and have overcome specific pain points, individuals can cultivate a network of support leverage diverse perspectives and champion inclusivity within their industries.

Strategies For Building Cross-Cultural Mentorship Relationships

Within Sreekanth's transformative journey under Bhagyashree's mentorship, they encountered a specific pain point, problem and subsequently developed effective solutions centered around building cross-cultural mentorship relationships.

Sreekanth's pain point stemmed from the challenges of bridging cultural gaps and effectively connecting with mentors from different backgrounds. As an individual navigating a diverse and multicultural landscape, he felt the need for strategies that would foster understanding, empathy and effective communication across cultural boundaries. Recognizing the significance of this pain point, Bhagyashree guided Sreekanth in developing strategies to build cross-cultural mentorship relationships. Together, they explored effective approaches that would enable Sreekanth to connect with mentors from diverse backgrounds and bridge the gaps in cultural understanding.

The first solution they implemented was the cultivation of cultural competence. Bhagyashree emphasized the importance of learning about different cultures, customs, and traditions. Sreekanth engaged in self-education, seeking resources, books, and documentaries that provided insights into various cultural backgrounds. This enabled him to approach cross-cultural mentorship relationships with an open mind and a genuine desire to understand and appreciate different perspectives. Furthermore, Sreekanth actively sought out opportunities to engage in cross-cultural experiences. He attended cultural festivals, volunteered in diverse communities, and participated in cross-cultural exchange programs. Through these immersive experiences, he developed a deeper understanding of cultural nuances and enhanced his ability to connect with individuals from diverse backgrounds. Bhagyashree also emphasized the importance of effective communication in cross-cultural mentorship relationships. She encouraged Sreekanth to adopt active listening skills, ensuring that he truly understood the perspectives, challenges, and aspirations of his mentors. Sreekanth honed his ability to ask thoughtful questions, seek clarification, and engage in empathetic conversations that transcended cultural differences.

To address the problem of building cross-cultural mentorship relationships, Bhagyashree introduced Sreekanth to the concept of mentoring circles or group mentoring. They explored opportunities where individuals from diverse backgrounds could come together and engage in mentorship activities. This provided a platform for mentors and mentees to learn from each other's experiences, fostering cross-cultural understanding and creating a supportive community. In addition, Bhagyashree encouraged Sreekanth to seek mentors who valued diversity and inclusivity. They actively sought out mentors who were known for their cross-cultural competence and ability to effectively navigate diverse environments. By connecting with mentors who were receptive to cross-cultural mentorship, Sreekanth fostered relationships based on mutual respect, understanding, and growth. Through the implementation of these strategies, Sreekanth successfully built meaningful cross-cultural mentorship relationships. He learned to navigate cultural differences with sensitivity and adapt his communication style to ensure effective engagement. These relationships not only provided him with invaluable guidance but also enriched his perspective, broadened his horizons, and facilitated personal and professional growth.

Sreekanth's ability to build cross-cultural mentorship relationships became a source of inspiration for others within his community. Witnessing his success, individuals from diverse backgrounds were encouraged to actively seek out mentors who could provide guidance and support, regardless of cultural differences. In the realm of mentorship and diversity, Sreekanth's journey serves as a testament to the power of implementing strategies for building cross-cultural mentorship relationships. By cultivating cultural competence, practicing effective communication, seeking mentors who value diversity, and engaging in mentorship circles, individuals can bridge cultural gaps, foster understanding, and create mutually beneficial mentorship relationships. The story of Sreekanth's cross-cultural mentorship journey exemplifies the transformative impact that can be achieved when individuals commit to building relationships that transcend cultural boundaries.

Through Bhagyashree's guidance, Sreekanth not only developed the skills necessary to navigate cross-cultural mentorship but also became an advocate for building inclusive and culturally sensitive mentorship networks within his community.

Addressing Potential Barriers To Effective Cross-Cultural Mentorship

Within Sreekanth's transformative journey under Bhagyashree's mentorship, they encountered a specific problem, pain, and developed effective solutions to address potential barriers in establishing effective cross-cultural mentorship relationships.

The problem they faced revolved around potential barriers that could hinder the effectiveness of cross-cultural mentorship. Sreekanth recognized that differences in communication styles, cultural norms, and implicit biases could impede meaningful connections and hinder the exchange of knowledge and guidance. Sreekanth experienced a pain point stemming from the fear of miscommunication and misunderstanding when engaging with mentors from different cultural backgrounds. He realized the importance of addressing these barriers to establish successful cross-cultural mentorship relationships and maximize the benefits of diverse perspectives. To address these potential barriers, Bhagyashree guided Sreekanth in implementing specific solutions. They focused on cultivating cultural intelligence and fostering open dialogue to mitigate misunderstandings and build strong cross-cultural connections.

The first solution they implemented was fostering cultural intelligence. Bhagyashree emphasized the importance of learning about cultural norms, values, and communication styles of mentors from different backgrounds. Sreekanth engaged in research, reading, and actively seeking out resources that provided insights into various cultural nuances. This enabled him to approach cross-cultural mentorship relationships with empathy and understanding. Bhagyashree also encouraged Sreekanth to engage in open dialogue with his mentors about potential cultural barriers. They discussed the importance of addressing misconceptions, biases, and stereotypes upfront, ensuring a foundation of trust and mutual respect. Sreekanth learned to ask thoughtful questions, seek clarifications, and approach conversations with curiosity and respect for his mentors' cultural backgrounds. To further address potential barriers, Sreekanth actively sought out mentors who embraced cultural diversity and were open to engaging in cross-cultural mentorship. Bhagyashree guided him in identifying mentors who demonstrated cultural sensitivity and had experience navigating diverse environments. These mentors not only provided valuable guidance but also acted as advocates for cross-cultural collaboration.

Another solution they implemented was creating a safe and inclusive space for cross-cultural mentorship. Sreekanth and Bhagyashree emphasized the importance of fostering an environment where mentors and mentees felt comfortable discussing cultural differences and potential challenges. They encouraged open and honest conversations, allowing for a deeper understanding of each other's perspectives and experiences. By addressing potential barriers, Sreekanth and Bhagyashree ensured that their cross-cultural mentorship relationships were effective and impactful. They acknowledged the importance of actively seeking to understand cultural differences and leveraging them as strengths, rather than allowing them to become barriers to connection and growth. Sreekanth's commitment to understanding and addressing potential barriers in cross-cultural mentorship became a catalyst for positive change within his community. He shared his experiences and strategies with others, encouraging them to approach cross-cultural mentorship with an open mind, cultural intelligence, and a willingness to engage in meaningful conversations. Through Bhagyashree's guidance, Sreekanth exemplified the power of addressing potential barriers in cross-cultural mentorship. By fostering cultural intelligence, open dialogue, and creating a safe space, individuals can build strong and effective cross-cultural mentorship relationships that transcend cultural differences and drive personal and professional growth.

In the realm of mentorship and diversity, Sreekanth's story serves as an inspiration for individuals striving to overcome potential barriers in cross-cultural mentorship. By actively addressing these challenges, individuals can build meaningful connections, leverage diverse perspectives, and create a more inclusive and empowering mentorship ecosystem.

The Benefits Of Diverse Mentorship Relationships

Within Sreekanth's transformative journey under Bhagyashree's mentorship, they recognized the benefits that diverse mentorship relationships could bring. They encountered a specific problem, pain, and developed effective solutions that emphasized the advantages of embracing diversity in mentorship.

The problem they faced stemmed from the lack of exposure to diverse perspectives and experiences. Sreekanth realized that limiting his mentorship relationships to individuals from similar backgrounds could result in a narrow worldview and missed opportunities for growth. He yearned for a solution that would allow him to harness the benefits of diverse mentorship relationships. Sreekanth experienced a pain point in feeling disconnected from a broader range of experiences and insights that mentors from diverse backgrounds could offer. He recognized that without embracing diversity in mentorship, he might overlook valuable perspectives and innovative approaches to problem-solving. To address this, Bhagyashree guided Sreekanth in implementing a solution that emphasized the benefits of diverse mentorship relationships. They actively sought out mentors from various backgrounds, fostering an environment that celebrated diversity and inclusivity.

By engaging with mentors from diverse backgrounds, Sreekanth discovered the tremendous benefits that emerged from these relationships. He gained access to a wealth of diverse perspectives, ideas, and approaches to challenges. The range of experiences shared by his mentors broadened his horizons and pushed him to think outside the box. The solution they implemented also encouraged Sreekanth to challenge his own assumptions and biases. Bhagyashree emphasized the importance of actively seeking out mentors who would provide differing viewpoints and perspectives. Sreekanth learned to value and appreciate the unique insights that mentors from diverse backgrounds brought to the table, fostering personal growth and expanding his understanding of the world. Through these diverse mentorship relationships, Sreekanth developed cultural competence, allowing him to navigate and communicate effectively across different backgrounds. The exposure to diverse perspectives not only enriched his personal growth but also enhanced his problem-solving skills and decision-making abilities. He began to recognize the power of collaboration and the strength that lies in embracing diversity.

The benefits of diverse mentorship relationships extended beyond Sreekanth's personal growth. As he applied the insights and knowledge gained from his mentors, he became an advocate for inclusivity and diversity within his community. Sreekanth's own success and empowerment inspired him to actively support and mentor individuals from underrepresented backgrounds, creating a positive ripple effect within his community. Sreekanth's journey exemplified the advantages of embracing diversity in mentorship relationships. By engaging with mentors from different backgrounds, he gained access to a multitude of perspectives, ideas, and approaches. The exposure to diverse experiences broadened his horizons and propelled him towards greater innovation and success. In the realm of mentorship and diversity, Sreekanth's story serves as a powerful reminder of the transformative benefits that come from embracing diverse mentorship relationships. By actively seeking out mentors from different backgrounds, individuals can expand their own perspectives, challenge biases, and foster a more inclusive and equitable society. Through Bhagyashree's guidance, Sreekanth unlocked the potential of diverse mentorship relationships, reaping the numerous benefits that arise from embracing diversity. His journey inspired others to recognize the advantages of diverse mentorship, sparking a wave of positive change within mentorship ecosystems and driving collective growth and innovation.

KEY TAKEAWAYS

CHAPTER 12: MENTORSHIP AND DIVERSITY: HOW TO FIND AND ENGAGE MENTORS FROM DIFFERENT BACKGROUNDS AND PERSPECTIVES

The Importance of Diversity In Mentorship Relationships

Sreekanth's transformative journey with Bhagyashree highlights the power of diverse mentorship relationships in addressing pain points, overcoming obstacles, and fostering personal growth. By embracing diversity, seeking mentors who understand unique challenges, and fostering an inclusive environment, individuals can unlock their full potential and pave the way for success.

How To Find Mentors From Different Backgrounds And Perspectives

Sreekanth's journey showcases the importance of seeking mentors from diverse backgrounds and perspectives, specifically those who have overcome barriers faced by marginalized communities. By connecting with mentors who understand unique challenges, individuals can gain invaluable guidance, empower themselves, and become agents of change in their industries, promoting inclusivity and breaking down systemic barriers.

Strategies For Building Cross-Cultural Mentorship Relationships

Sreekanth's journey highlights the importance of strategies for building cross-cultural mentorship relationships. By cultivating cultural competence, practicing effective communication, seeking mentors who value diversity, and engaging in mentorship circles, individuals can bridge cultural gaps, foster understanding, and create mutually beneficial mentorship relationships.

Addressing Potential Barriers To Effective Cross-Cultural Mentorship

Sreekanth's journey emphasizes the importance of addressing potential barriers to effective cross-cultural mentorship. By fostering cultural intelligence, promoting open dialogue, and creating a safe and inclusive space, individuals can overcome cultural differences and build strong and impactful cross-cultural mentorship relationships.

The Benefits of Diverse Mentorship Relationships

Sreekanth's journey highlights the significant benefits of diverse mentorship relationships. By engaging with mentors from different backgrounds, individuals gain access to diverse perspectives, innovative ideas, and enhanced problem-solving skills. Embracing diversity in mentorship fosters personal growth, inspires advocacy for inclusivity, and drives collective growth and innovation within communities.

CHAPTER 13: MENTORSHIP BEYOND BOUNDARIES: HOW TO ESTABLISH AND MAINTAIN LONG-DISTANCE MENTOR-MENTEE RELATIONSHIPS

> *"A leader takes people where they want to go. A great leader takes people where they don't necessarily want to go, but ought to be."* –
> Rosalynn Carter

Let me share with you a powerful and inspirational story that happened in New York City, USA on March-2011, that embodies the essence of mentoring beyond boundaries. David, a passionate individual with a burning desire to establish his coaching business specializing in long-distance mentorship. David had been on a challenging journey, struggling to find a mentor who could meet his specific expectations and gid him to success in the real of long-distance coaching. After months of searching and connecting with numerous mentors, David finally came across Lisa, a renowned mentor with an exceptional track record of mentoring individuals in long-distance coaching. Lisa had the ability to effectively deal with the unique challenges and empower and inspire individuals located far away. Initially, Lisa was hesitant to take on David as a mentee. She wanted to ensure that David was truly committed and dedicated to his coaching business. Lisa decided to test David's perseverance and consistency before committing to mentorship full. She gave David a complex task to complete within a strict deadline. The task required David to develop a comprehensive coaching program specifically tailored for long-distance clients. David was determined to prove his worth and embraced the challenge wholeheartedly.

As David delved into the task, he encountered numerous obstacles. He struggled to figure out the most effective ways to structure his coaching program and provide valuable support to clients who were physically distant. Despite his initial setbacks, David refused to give up. He sought guidance from various resources, experimented with different strategies, and persevered through countless iterations. Unfortunately, David's efforts fell short, and he failed to meet the deadline set by Lisa. However, Lisa, who had been observing David's journey closely, was deeply impressed by his unwavering focus, dedication, consistency, and persistence. She recognized that David possessed the qualities of a resilient and determined individual who could benefit greatly from her mentorship. Seeing David's unwavering commitment and recognizing his potential, Lisa's perspective shifted. She decided to take David under her wing and mentor him in establishing a powerful and successful long-distance coaching business. Their mentorship journey extended beyond geographical boundaries as they navigate challenges and achieve extraordinary success.

One of the pain points David faced in his coaching business was how to effectively connect and support clients who were physically distant. Through Lisa's guidance, he discovered innovative technological solutions, such as video conferencing, online learning platforms, and collaborative tools. These tools enabled him to bridge the physical gap, connect deeply with his clients, and provide personalized coaching experiences that exceeded their expectations. With Lisa's mentorship, David developed a strong business strategy, refined his coaching techniques, and established a powerful online presence. He leveraged social media platforms, webinars, and networking events to attract clients from around the globe. David's reputation as a top long-distance coach grew, and he started receiving inquiries and testimonials from clients who experienced incredible transformation under his guidance. The long-distance mentorship provided by Lisa enabled David to overcome the challenges and limitations imposed by physical boundaries. He not only built a successful coaching business but also created a positive impact on the lives of individuals who previously struggled to access coaching support due to geographical limitations. The story exemplifies how long-distance mentorship can help individuals like David overcome obstacles, find innovative solutions to unique challenges, and achieve success beyond their wildest dreams. By embracing the power of technology, maintaining unwavering dedication, and seeking guidance from supportive mentors, the possibilities of success in long-distance coaching become boundless.

David's journey showcases the transformative power of mentoring beyond boundaries. Through Lisa's guidance and support, he was able to navigate the complexities of long-distance coaching, adapt to new technologies and create a thriving business that transcended geographical limitations. Their mentorship went beyond the traditional mentor-mentee relationship, as they formed a strong bond based on trust, respect, and shared goals. By leveraging technology, David was able to connect with clients from all corners of the world, breaking down barriers and offering his expertise to those who previously had limited access to coaching. He developed a reputation for providing impactful and transformative coaching experiences, empowering individuals to overcome their challenges, unlock their full potential and achieve extraordinary success. Through Lisa's mentorship, David gained not only practical knowledge and strategies but also the confidence to thrive in the competitive world of long-distance coaching. He learned to embrace his unique strengths, build a strong personal brand, and consistently deliver exceptional value to his clients. The long-distance mentorship journey pushed David beyond his comfort zone, allowing him to discover new ways of connecting, supporting and empowering his clients.

The success David achieved was far beyond his initial expectations. His coaching business flourished, attracting clients from diverse backgrounds and geographical locations. He became a sought-after thought leader in the field of long-distance coaching, speaking at conferences and sharing his expertise through publications and interviews. Furthermore, David's success inspired others to embrace the possibilities of long-distance mentorship. Many aspiring coaches who faced similar challenges as David sought his guidance and mentorship, further expanding his impact and reach. This real-time story of mentoring beyond boundaries exemplifies the profound impact that a dedicated mentor can have on an individual's success. By finding the right mentor who understands the intricacies of their chosen niche and possesses the knowledge, experience, and willingness to support from a distance, individuals like David can surpass their own expectations and achieve remarkable results. The story of David and Lisa serves as a powerful reminder that no obstacle is insurmountable when guided by a mentor who believes in your potential and supports your growth. Long-distance mentorship opens doors to opportunities that may have seemed out of reach, enabling individuals to break free from limitations and achieve greatness in their chosen fields.

The Benefits of Long-Distance Mentorship Relationships

One of the key benefits David gained from his long-distance mentorship relationship with Lisa was the ability to tap into a wealth of knowledge and experience. Lisa had a proven track record in mentoring individuals in the field of long-distance coaching, and her expertise proved invaluable to David's journey. Through their mentorship, David gained insights into effective coaching techniques, client engagement strategies, and business development tactics specific to the challenges of coaching clients from afar. One pain point that David encountered was the difficulty of establishing meaningful connections with clients who were physically distant. This posed a challenge as building trust and rapport are essential elements of any coaching relationship. Recognizing this hurdle, Lisa shared her own experiences and provided David with a range of practical solutions. Together, they explored the power of video conferencing and virtual platforms, allowing David to conduct coaching sessions that replicated the personal interaction and connection of in-person meetings. By utilizing these tools, David was able to bridge the distance and establish deep, impactful relationships with his clients.

Another problem David faced was the potential isolation and lack of a supportive network that can arise when working with long-distance clients. Lisa understood the importance of community and introduced David to her network where like-minded professionals discuss their problems and work together for solutions through online coaching forums and communities. By connecting with others who shared his passion for long-distance coaching, David gained a supportive network that provided not only camaraderie but also valuable and collaboration opportunities. Lisa's mentorship also helped David navigate the ever-evolving landscape of technology. As an aspiring coach with a focus on long-distance mentoring, it was crucial for David to stay up to date with the latest advancements in virtual communication, collaboration tools, and online marketing platforms, Lisa guided him through the process of selecting and utilizing the most effective technology solutions for his coaching business. With her guidance, David gained a competitive edge, leveraging technology to streamline his operations, expand his reach, and deliver exceptional value to his clients. A significant benefit that David derived from his mentorship with Lisa was the development of a strong mindset and unwavering self-belief. Lisa empowered him to embrace his unique strengths and recognize the value he brought to his clients. Through her guidance, David learned to overcome self-doubt and imposter syndrome, allowing him to show up confidently and authentically in his coaching sessions. This newfound self-assurance translated into better client outcomes, increased client satisfaction, and ultimately a thriving coaching practice.

Moreover, Lisa's mentorship provided David with ongoing accountability and support. They established regular check-in meetings and set milestones to measure David's progress. Lisa held him accountable for his goals, providing constructive feedback and celebrating his achievements along the way. This level of mentorship ensured that David remained focused, motivated, and committed to his personal and professional growth. Through Lisa's mentorship relationships, David not only achieved success in his coaching business but also experienced personal growth and fulfillment. He developed a strong sense of purpose, knowing that he was making a positive impact on the lives of his clients, even from a distance. The long-distance mentorship relationship empowered David to realize his full potential, help him become the coach he aspired to be. The benefits of long-distance mentorship relationships are vast and far-reaching. David's mentorship journey with Lisa showcased the power of having a mentor who understands the unique challenges of long-distance coaching and can provide guidance, support, and expertise tailored to those circumstances. From bridging the physical distance with technology to cultivating a supportive network and developing a strong mindset, David's mentorship relationships with Lisa paved the way for his exceptional success in the field of long-distance coaching.

Strategies For Finding and Connecting With Mentors Outside of Your Geographic Area

Finding and connecting with mentors outside of one's geographic area can be a transformative step in personal and professional growth. David's journey of long-distance mentorship with Lisa not only exemplifies the power of such relationships but also provides insights into effective strategies for connecting with mentors beyond physical boundaries. One pain point that David initially faced was the challenge of identifying potential mentors who specialized in long-distance coaching and could cater to his specific needs. Recognizing this hurdle, Lisa shared her expertise on how to locate mentors outside of his immediate geographic area. She encouraged David to leverage online platforms, professional networking sites, and coaching communities to connect with individuals who had experience in long-distance coaching. Lisa guided David in conducting thorough research to identify potential mentors who aligned with his coaching niche and had a track record of success in mentoring individuals remotely. Through this process, David discovered a multitude of professionals who were willing to share their knowledge and insights, despite the physical distance between them. Once potential mentors were identified, the next step was to initiate meaningful connections. Lisa emphasized the importance of crafting personalized and compelling messages that demonstrated genuine interest and a clear understanding of the potential mentor's expertise. By showcasing his passion, commitment, and alignment with the mentor's values, David increased the likelihood of receiving a positive response.

Lisa also guided David on the art of building and nurturing relationships with potential mentors. She encouraged him to approach the mentorship process with a mindset of mutual value exchange. David learned to identify ways in which he could contribute to the mentor's journey, such as sharing his own experiences, providing feedback on their work, or offering assistance in specific areas where he had expertise. This approach helped David establish authentic and reciprocal connections, laying the foundation for meaningful mentorship relationships. To further enhance his chances of securing a mentor, Lisa advised David to be persistent and resilient. She emphasized the importance of consistently following up on initial communications, demonstrating his dedication and commitment to the mentorship process. David learned to respect the mentor's time and boundaries while expressing his sincere desire for guidance and growth. When David finally connected with Lisa, his ideal mentor, he encountered a significant challenge: convincing her of his dedication and consistency. Lisa initially hesitated to take on new mentees due to the high standards she maintained in her mentorship relationships. To overcome this hurdle, Lisa provided David with a difficult task that tested his perseverance and commitment. While David faced initial setbacks and struggled to complete the task within the given deadline, his unwavering focus, dedication, consistency, and persistence left a lasting impression on Lisa. She recognized his potential and shifted her perspective, ultimately agreeing to mentor him. This pivotal moment demonstrated the importance of staying committed and resilient, even in the face of challenges, to gain the trust and confidence of potential mentors. Throughout their mentorship relationship, David continued to implement the strategies shared by Lisa. He actively sought out opportunities to provide value to her and contribute to their relationship, fostering a strong bond of mutual respect and admiration. This approach ensured the longevity and depth of their mentorship. The strategies for finding and connecting with mentors outside of one's geographic area are vital for accessing valuable guidance and support. David's journey with Lisa showcased the power of conducting thorough research, crafting personalized messages, building authentic relationships, and demonstrating dedication and resilience. By implementing these strategies, David not only overcame initial barriers but also secured a mentor who played a pivotal role in his long-distance coaching success. The experience highlighted the transformative potential of mentorship relationships that transcend physical boundaries, enabling individuals to reach new heights in their personal and professional endeavors.

Best Practices for Maintaining Long-Distance Mentorship Relationships

Maintaining a long-distance mentorship relationship requires effective communication, trust, and commitment from both the mentee and the mentor. David and Lisa's journey exemplifies the best practices they employed to ensure the success and longevity of their mentorship relationship, overcoming potential challenges that arise when physically distant. One pain point that David initially encountered was the potential difficulty of maintaining a strong connection and regular communication with Lisa due to the geographical distance between them. To address this challenge, David and Lisa established a framework of best practices to maintain open and effective communication. They prioritized regular virtual meetings using video conferencing platforms. These meetings provided an opportunity for face-to-face interaction, allowing David to receive direct feedback, guidance, and support from Lisa. They scheduled these meetings in advance, ensuring that both parties were prepared and committed to the mentorship discussions.

Additionally, David and Lisa implemented a consistent and reliable means of communication outside of the scheduled meetings. They leveraged email, instant messaging platforms, and shared online documents to stay connected in between their virtual sessions. This allowed for ongoing updates, progress reports, and the exchange of resources and materials relevant to David's coaching business. Another challenge David faced was the potential for miscommunication or misunderstandings due to the lack of non-verbal cues that come with in-person interactions. Lisa guided David on effective communication strategies, emphasizing the importance of active listening and clarity in conveying thoughts and ideas. David learned to be proactive in seeking clarification whenever needed, ensuring that their communication remained precise and aligned. To further strengthen their mentorship relationship, David and Lisa fostered trust and transparency. They established a safe and non-judgmental space for open dialogue, where David felt comfortable sharing his aspirations, challenges, and progress. Lisa provided constructive feedback, encouragement, and guidance while maintaining a supportive environment that nurtured David's growth. Recognizing the value of accountability, David and Lisa implemented goal-setting practices. They collaboratively set short-term and long-term goals, ensuring that these objectives aligned with David's vision and aspirations. Lisa held David accountable by reviewing his progress regularly and providing guidance to keep him on track. These goal-setting practices provided structure and focus to their mentorship relationship, empowering David to continually strive for excellence.

Furthermore, David and Lisa embraced a continuous learning mindset. They explored opportunities to expand their knowledge and expertise in the field of long-distance coaching. Lisa recommended relevant books, articles, and industry resources to David, enabling him to stay updated with the latest trends and best practices. This commitment to ongoing learning fueled their discussions and allowed for a deep dive into relevant topics during their mentorship sessions. Maintaining a long-distance mentorship relationship requires adaptability and the willingness to leverage technology effectively. David and Lisa utilized online collaboration tools to streamline their mentorship processes, such as sharing and co-editing documents, tracking progress, and setting reminders for key milestones. These digital tools not only facilitated efficient communication but also created a shared virtual workspace that enhanced their productivity and organization. David and Lisa employed various best practices to successfully maintain their long-distance mentorship relationship. Through open and effective communication, building trust, setting goals, and embracing continuous learning, they created a strong foundation for their mentorship journey. By employing these best practices, David not only overcame the challenges of physical distance but also thrived in his long-distance coaching business under Lisa's expert guidance. Their commitment to maintaining a meaningful mentorship relationship serves as a testament to the transformative power of long-distance mentorship.

Addressing Common Challenges in Long-Distance Mentorship Relationships

Long-distance mentorship relationships, while incredibly rewarding, can present unique challenges for both mentees and mentors. David and Lisa's journey serves as an example of their ability to address and overcome common challenges, ultimately achieving success in their long-distance mentorship relationship. One challenge David faced was the potential difficulty of establishing a strong connection with Lisa due to the physical distance between them. As a mentee, he desired a mentor who would understand his specific needs as a long-distance coach. Lisa recognized this challenge and actively worked towards creating a meaningful connection with David. To overcome this hurdle, Lisa dedicated time to understand David's background, goals, and aspirations. Through open and transparent communication, she fostered an environment where David felt valued and understood. By demonstrating her genuine interest and commitment to his success, Lisa solidified their connection and built a foundation of trust and mutual respect.

On the other hand, Lisa faced the challenge of effectively understanding David's unique circumstances and tailoring her guidance to his specific needs. As a mentor, she recognized the importance of acknowledging and addressing these challenges to ensure that David received the support necessary for his growth. Lisa actively sought to understand the pain points David faced in his long-distance coaching journey. Through deep listening and empathetic discussions, she encouraged David to express his concerns and challenges openly. By gaining a comprehensive understanding of David's situation, Lisa was able to provide tailored solutions and guidance specific to his long-distance coaching business. One significant pain point for David was the potential isolation and lack of a supportive network as a long-distance coach. Recognizing this challenge, Lisa encouraged David to connect with like-minded professionals and join online coaching communities. This approach allowed him to build relationships, seek advice, and find support from individuals who shared similar experiences and challenges. Lisa also helped David develop strategies to overcome communication barriers that can arise in long-distance mentorship relationships. She emphasized the importance of clear and concise communication, ensuring that both parties were aligned in their understanding of expectations, goals, and progress. By fostering effective communication channels, David and Lisa overcame potential misunderstandings and maintained a strong mentorship relationship.

Another common challenge in long-distance mentorship relationships is the potential lack of accountability and commitment. To address this, David and Lisa established a framework for goal setting and progress tracking. They set clear milestones, discussed action plans, and scheduled regular check-ins to review progress and provide guidance. This practice ensured that David remained accountable and committed to his personal and professional growth. Moreover, both David and Lisa had to navigate the complexities of technology and virtual communication. They encountered occasional technical glitches and connectivity issues that could have disrupted their mentorship relationship. However, they approached these challenges with adaptability and resourcefulness. They identified alternative communication platforms, explored backup options, and remained patient and understanding during any technical difficulties that arose. By addressing these common challenges head-on, David and Lisa demonstrated their commitment to the success of their long-distance mentorship relationship. Through effective communication, empathy, tailored guidance, and a focus on accountability, they overcame obstacles that could have hindered their progress. Their willingness to adapt, support, and understand each other's circumstances paved the way for a fruitful and impactful mentorship experience. David and Lisa's journey showcases their ability to address and overcome common challenges in long-distance mentorship relationships. By actively working to establish a strong connection, tailoring guidance to specific needs, fostering effective communication, maintaining accountability, and navigating technological hurdles, they forged a successful mentorship relationship that enabled David to thrive in his long-distance coaching business. This serves as a testament to the power of addressing challenges proactively and collaboratively to achieve success in long-distance mentorship relationships.

The Role of Technology in Maintaining Long-Distance Mentorship Relationships

Technology plays a pivotal role in maintaining successful long-distance mentorship relationships. David and Lisa leveraged various technological tools to bridge the physical gap and ensure effective communication and collaboration. However, they also encountered challenges along the way and worked together to overcome them, ultimately achieving remarkable success in their mentorship journey. One pain point that David initially faced was the need to establish a reliable and efficient means of communication with Lisa despite the distance between them. Recognizing the importance of timely and seamless communication, they utilized video conferencing platforms such as Zoom and Skype for their virtual meetings. These platforms allowed them to have face-to-face conversations, ensuring a deeper connection and the ability to interpret non-verbal cues. However, technical challenges, such as poor internet connectivity or audio/video disruptions, posed occasional obstacles during their virtual meetings. To overcome these challenges, David and Lisa implemented strategies to mitigate potential disruptions. They agreed to conduct virtual meetings during periods of stable internet connectivity, allowing for uninterrupted discussions. In cases where technical issues arose, they remained patient, understanding that occasional disruptions were an inevitable part of the virtual landscape. They also explored alternative communication channels, such as phone calls or email, as backup options when needed.

Another pain point David faced was the potential difficulty of sharing documents, resources, and progress updates with Lisa. Recognizing the importance of collaborative work, they leveraged cloud-based storage and sharing platforms, such as Google Drive or Dropbox. These tools provided a centralized space where David could upload and share files, allowing Lisa to access and review them at her convenience. This streamlined their collaboration, ensuring efficient feedback and progress tracking. However, the challenge of data security and privacy in sharing sensitive information posed concerns for both David and Lisa. To address this, they implemented security measures such as password protection and encryption for their shared documents. They also maintained open and transparent communication regarding the importance of data privacy and took steps to ensure the confidentiality and integrity of their shared information. Additionally, David and Lisa recognized the significance of ongoing engagement and communication outside of their scheduled virtual meetings. They utilized instant messaging platforms, such as Slack or Microsoft Teams, to maintain real-time communication, share quick updates, and address any urgent questions or concerns. These platforms facilitated continuous connection and allowed for timely support, fostering a sense of availability and responsiveness in their mentorship relationship. However, the potential distraction and overwhelm caused by constant notifications and digital noise posed a challenge. To overcome this, David and Lisa established clear boundaries and communicated their availability preferences. They agreed on designated communication hours, during which they would be responsive to each other's messages and requests. This created a balance between staying connected and allowing for focused work and personal time.

David and Lisa effectively utilized technology to maintain their long-distance mentorship relationship. By leveraging video conferencing, cloud storage, instant messaging, and other collaborative tools, they ensured effective communication, seamless collaboration, and timely support. Despite occasional technical challenges and concerns regarding data security, they remained adaptable, resourceful, and patient. Their ability to navigate and overcome these technological hurdles was instrumental in the success of their mentorship journey. The story of David and Lisa exemplifies the transformative power of technology in maintaining long-distance mentorship relationships. By harnessing the capabilities of digital tools and platforms, individuals can connect, collaborate, and support each other regardless of geographical barriers. The successful integration of technology by David and Lisa demonstrates that, with the right mindset and strategies, technology can serve as a powerful enabler for mentorship that transcends physical boundaries.

KEY TAKEAWAYS

CHAPTER 13: MENTORSHIP BEYOND BOUNDARIES: HOW TO ESTABLISH AND MAINTAIN LONG-DISTANCE MENTOR-MENTEE RELATIONSHIPS

The Benefits of Long-Distance Mentorship Relationships

The powerful and inspirational story of David and Lisa demonstrates the transformative power of long-distance mentorship. By embracing technology, maintaining unwavering dedication, and seeking guidance from a supportive mentor, individuals can overcome obstacles, find innovative solutions, and achieve success beyond their wildest dreams. Long-distance mentorship opens doors to opportunities that may have seemed out of reach, enabling individuals to break free from limitations and achieve greatness in their chosen fields.

Strategies For Finding and Connecting with Mentors Outside of Your Geographic Area

Finding and connecting with mentors outside of one's geographic area requires strategic research, personalized outreach, and a mindset of mutual value exchange. David's journey with Lisa demonstrates the importance of conducting thorough research to identify potential mentors, crafting compelling messages, building authentic relationships, and demonstrating dedication and resilience. By implementing these strategies, individuals can overcome barriers and secure mentors who can provide valuable guidance and support, even from a distance. The transformative potential of long-distance mentorship relationships becomes accessible when one proactively seeks connections beyond physical boundaries.

Best Practices for Maintaining Long-Distance Mentorship Relationships

Maintaining a successful long-distance mentorship relationship requires effective communication, trust, accountability, and a commitment to continuous learning. David and Lisa's journey demonstrates the best practices they employed, including regular virtual meetings, consistent communication channels, active listening, goal setting, and leveraging technology for collaboration. By implementing these best practices, mentees and mentors can overcome the challenges of distance and foster a strong and impactful mentorship relationship that fuels personal and professional growth.

Addressing Common Challenges in Long-Distance Mentorship Relationships

Addressing common challenges in long-distance mentorship relationships requires open communication, understanding, tailored guidance, accountability, and adaptability. David and Lisa's journey demonstrates the importance of establishing a strong connection, addressing unique circumstances, fostering effective communication, maintaining accountability, and navigating technological hurdles. By actively working together to overcome these challenges, mentees and mentors can forge successful and impactful long-distance mentorship relationships that support growth and success.

The Role of Technology in Maintaining Long-Distance Mentorship Relationships

Technology plays a crucial role in maintaining successful long-distance mentorship relationships by facilitating effective communication, collaboration, and support. David and Lisa utilized video conferencing, cloud storage, instant messaging, and other collaborative tools to bridge the physical gap and streamline their mentorship process. While they encountered technical challenges and concerns about data security, they remained adaptable and implemented strategies to mitigate disruptions and ensure privacy. Their story showcases the transformative power of technology in breaking down geographical barriers and enabling meaningful mentorship connections.

May I ask you for a small favor?

I hope this note finds you well and filled with joy. As an author, I couldn't be more grateful for your decision to embark on this literary journey with me by reading my book, **"The Power of Mentor – Volume II"** Your time and attention are incredibly valuable, and I am deeply touched that you chose to invest them in exploring the world I created within these pages.

Writing this book was a labor of love and knowing that it has touched your heart and mind means the world to me. Your support and encouragement have inspired me beyond measure, reminding me of the profound impact words can have on our lives and how they can forge genuine connections between strangers.

If it's not too much to ask, I would be profoundly grateful if you could take a few moments to share your thoughts on Amazon through a review. Your feedback will not only help potential readers discover the book but will also guide me in my future writing endeavors. Your honest opinion is priceless.

To put it straight – **Reviews are the lifeblood for any author.**

Additionally, I wholeheartedly recommend **"The Ultimate Leadership in You"** & **"The Power of Mentor Volume I"** to anyone seeking to unlock their true leadership potential and make a profound impact on their life and the lives of those around them. Your personal recommendation could be the catalyst for transformation in the lives of those who need it most.

Remember, a heartfelt review has the power to touch the lives of countless readers, guiding them to a story that could make a difference in their lives too.

Once again, thank you for joining me on this literary adventure. Your presence in my journey is a cherished gift, and I am forever grateful for the opportunity to connect with you through my words.

Cheers

Sreekanth Ganeshi

My Other Book

Click Here to Buy Now   Click Here to Buy Now

Follow us on Social Media

https://www.facebook.com/SreekanthGaneshi

https://www.facebook.com/groups/sreekanthganeshi

https://twitter.com/1Sreekanth_G

https://www.linkedin.com/in/sreekanthganeshi/

https://www.instagram.com/sreekanthganeshi/

Notes

Notes

Notes

Printed in Poland
by Amazon Fulfillment
Poland Sp. z o.o., Wrocław